ISBN 978-1-332-02207-6
PIBN 10270322

This book is a reproduction of an important historical work. Forgotten Books uses
state-of-the-art technology to digitally reconstruct the work, preserving the original format
whilst repairing imperfections present in the aged copy. In rare cases, an imperfection in
the original, such as a blemish or missing page, may be replicated in our edition. We do,
however, repair the vast majority of imperfections successfully; any imperfections that
remain are intentionally left to preserve the state of such historical works.

1 MONTH OF
FREE
READING

at

www.ForgottenBooks.com

By purchasing this book you are eligible for one month membership to ForgottenBooks.com, giving you unlimited access to our entire collection of over 700,000 titles via our web site and mobile apps.

To claim your free month visit: www.forgottenbooks.com/free270322

English
Français
Deutsche
Italiano
Español
Português

www.forgottenbooks.com

Mythology Photography **Fiction**
Fishing Christianity **Art** Cooking
Essays Buddhism Freemasonry
Medicine **Biology** Music **Ancient
Egypt** Evolution Carpentry Physics
Dance Geology **Mathematics** Fitness
Shakespeare **Folklore** Yoga Marketing
Confidence Immortality Biographies
Poetry **Psychology** Witchcraft
Electronics Chemistry History **Law**
Accounting **Philosophy** Anthropology
Alchemy Drama Quantum Mechanics
Atheism Sexual Health **Ancient History**
Entrepreneurship Languages Sport
Paleontology Needlework Islam
Metaphysics Investment Archaeology
Parenting Statistics Criminology
Motivational

MASONRY DAMS

75640

FROM INCEPTION TO COMPLETION

INCLUDING

NUMEROUS FORMULÆ, FORMS OF SPECIFICATION AND TENDER, POCKET DIAGRAM OF FORCES, ETC.

For the use of Civil and Mining Engineers

BY

C. F. COURTNEY, M.Inst.C.E.

LATE ASSISTANT ENGINEER, FAIRBAIRN ENGINEERING CO.; ASSISTANT ENGINEER
TO THE CITY SURVEYOR OF MANCHESTER; ENGINEER-IN-CHIEF
THARSIS SULPHUR AND COPPER CO., SPAIN

LONDON

CROSBY LOCKWOOD AND SON

7 STATIONERS' HALL COURT, LUDGATE HILL

1897

BRADBURY, AGNEW, & CO. LD., PRINTERS,
LONDON AND TONBRIDGE.

PREFACE.

THE Author desires it to be understood that this little work does not pretend to furnish an exhaustive treatise on the Design and Construction of Masonry Dams, but merely to draw together briefly some necessary information, and to offer some assistance to the student in overcoming the many difficulties which are likely to present themselves in an undertaking of merit and importance.

The constant requirement of large bodies of water for mining works (works of that kind being now in active development and extension in all parts of the world—up-country mostly, oftentimes removed from opportunities of obtaining complete information as to water storage, &c., put together in a compact form) has induced the Author to write the following pages, in the hope that they may be of assistance particularly to those whose work is carried on away from the centres of practical information. Though presented by him with diffidence, it is hoped that the data herein condensed may assist his fellow mining engineers in a few detailed difficulties and dangers.

The graphic system has been adopted in indicating the limits of pressure in the dam, being considered preferable to that of calculation. Any error in the use of simple graphic methods, instead of elaborate calculations, is at once proclaimed by the polygon of forces refusing to close ; whereas a figure or a decimal point in calculations wrongly placed, but in every other way correct, may easily lead to extremely erroneous results, and possibly endanger the stability of the structure to which the problem relates. Full detailed calculations of stability are, however, given, as it may be desirable to employ both methods.

Carefully reading a book will not make an engineer ; and it may be as well to remind those who are too apt to think otherwise that twenty or thirty years' experience cannot be obtained in that easy and rapid manner; and further, that although an effort is made in the following pages to clearly embrace the salient features in the construction of Masonry Dams from conception to finish, there is still left ample room for experience to affect the cost of construction. Moreover, to appreciate the delicacies of pressure, stability, and, above all, the dangers and risks of leakage, in such works, is vouchsafed only to those who have had long experience in engineering work.

Uniformity and completeness in a volume like the

present are only compatible with ample leisure, whilst the papers which form this little treatise were prepared during the spare evenings of active work.

It would be impossible to write such a book, or for the work to possess any practical value, without making frequent use of what has already been published; but where that has been done care has been taken, as far as possible, to acknowledge the source of information. Any omission in that respect is not due to intention.

C. F. COURTNEY.

Note.—In the absence abroad of Mr. C. F. Courtney, this volume has passed through the press without his personal supervision of the proof-sheets, and it has been a pleasure to me to undertake for him the friendly service of checking the proofs with his MSS.

HASTINGS C. DENT,
Assoc.-Memb. Inst.C.E.

London,
June, 1897.

CONTENTS.

———•———

CONTENTS.

APPENDIX III.

LIST OF ILLUSTRATIONS.

MASONRY DAMS.

SITE AND POSITION.

A SUBJECT of important consideration is the choice of situation, for, whether a dam be constructed for the retention of water for household consumption, irrigation, or manufacturing requirements, the site selected should be as free as possible from vegetation and great depth of soil, both being detrimental to cleanliness, purity of the water, and quantity given by the drainage area from the rainfall. A gradually rising area of great extent is preferable, so that the impounding space is large, with a dam of moderate height. Any area that has in past times become a lagoon or lake, and by the action of erosion cut an outlet forming a narrow neck for its discharge, is eminently suited, as invariably the upper and surrounding country is extensive, and insures an abundant supply of water from the rainfall. The sites are therefore restricted, and confined to such districts as are neither mountainous nor extremely flat, with sluggish watercourses. A position that commands an extensive drainage area, such as will yield more than the required quantity of water from the average yearly rainfall without the necessity of extension by the employment of catch drains, becomes at once a most favourable site, as in no case will the water brought into requisition by catch drains supply the same available percentage of the rainfall as that derived from the catchment area appertaining to the reservoir.   Should the area of the water retained in the reservoir by the dam be extensive, the evaporation will be considerable, and directly proportionate to the surface exposed to the sun's rays during dry seasons. The effects of evaporation must also be clearly borne in mind, as the total estimated to be retained for consumption will be seriously reduced.   In humid climates, such as England, the evaporation is not excessive, ranging between five and ten per cent. of the total retained, whilst in the South of Spain and Egypt thirty-three per cent. is no unusual loss.   From four to five feet depth of water from the surface of a reservoir is by no means an exceptional annual evaporation in those countries.   It is therefore only possible in such regions, with light rainfall and great evaporation, to obtain an abundant supply by either diverting a river to the site of the reservoir or selecting an immense drainage area that conveniently converges towards the works.

For simplification of calculating the available water obtained from a given area the following might with advantage be committed to memory : Every centimetre of rainfall is equal to one cubic metre of water when falling upon one hundred square metres of ground : which, when multiplied by two hundred and twenty, results in gallons. About fifty-five per cent. of this quantity will be found available when falling upon barren ground. Should, however, the annual rainfall be as low as twelve inches, and distributed over nine months of the year, the available quantity will not be more than twenty to twenty-five per cent.; if, on the other hand, between thirty and forty inches of rain fall during the same period, fifty to fifty-five per cent. will be obtained. Heavy storms will give as much as eighty to eighty-four per cent. of their water, there not being sufficient time for brushwood, heather, soil, or permeable detrimental strata to absorb the flood produced. Practically the whole of the water, therefore, runs rapidly into the watercourses.

Dew is also an element which affects an available supply, as in some countries it is very heavy, and if produced either from the moisture of the ground or condensed from the atmosphere it is still there, ready to assist the night or early morning rainfall to run freely into the valleys.

The evaporation from the sea on the shores of the Mediterranean is sufficient to yield about five times the quantity brought down by the watercourses. Messieurs Mariotte and Dausse ascertained that the annual quantity carried down by the Seine is not more than one-third of that supplied by the atmosphere to the country which it drains; the remaining two-thirds of the rain must, therefore, either be evaporated or absorbed by the vegetation. Carefully observing the foregoing features with study, and inquiring on the ground of an ultimately selected site, an estimate can be made of the probable annual reliable supply that would be given by the drainage area.

Unfortunately many are unacquainted with the metric system. The following table will therefore assist in the computation of the cubic feet of water received per acre, &c., from various rainfalls :—

Rainfall in inches ...	1	2	3	4	5	6	7	8
Cubic feet per acre...	3,630	7,260	10,890	14,520	18,150	21,780	25,410	29,040
Million cubic feet per square mile	$2\frac{1}{3}$	$4\frac{1}{2}$	7	$9\frac{1}{4}$	$11\frac{1}{2}$	14	$16\frac{1}{4}$	$18\frac{1}{2}$

Rainfall in inches ...	9	10	12	15	20	25	30
Cubic feet per acre...	32,670	36,300	43,560	54,450	72,600	90,750	108,900
Million cubic feet per square mile	21	$23\frac{1}{4}$	$27\frac{3}{4}$	$34\frac{3}{4}$	$46\frac{1}{2}$	58	$69\frac{3}{4}$

Rainfall in inches ...	40	50	60	70	80	90	100
Cubic feet per acre...	145,200	181,500	217,800	254,100	290,400	326,700	363,000
Million cubic feet per square mile	93	$116\frac{1}{4}$	$139\frac{1}{2}$	$162\frac{1}{2}$	$185\frac{3}{4}$	209	$232\frac{1}{4}$

The undermentioned multipliers will also be found serviceable :—Inches of rainfall × 2,323,200 = cubic feet per sq. mile. Inches of rainfall × 14½ = millions of gallons per sq. mile. Inches of rainfall × 3,630 = cubic feet per acre. One cubic foot × 6,235 = gallons. One cubic metre × 220 = gallons.

As the general health of a district may be seriously affected by the proximity of a large area of water in warm or semi-tropical climates, it is necessary to remark that all stagnant water is productive of malaria ; the sudden change of temperature in hot climates at sundown or sunrise causes a sickening and chilling miasma to rise from all low-lying ground, which is greatly intensified where water is retained. That which was intended, therefore, for the benefit of a district may become of doubtful advantage to the immediate neighbourhood, a persistent and insidious fever being the result of what is otherwise a magnificent and beneficent water supply.

In utilising a river as a permanent supply, calculations require to be made at various times in the year of its discharge before any definite opinion can be formed as to its minimum supply, and for this purpose a weir or notch board is the simplest and most convenient. This is very easily erected at small cost, consisting of a small dam, made by fixing planks across the river, and a notch cut upon the upper plank; a sill, over which the whole of the water of the stream or river is made to run, being formed by cutting a feather edge on the wood, or by fixing a thin brass plate. The opening must have perfectly vertical sides, and the sill be horizontal; the water approaching the weir should not have an appreciable velocity, or the calculation is more troublesome.

The formula that can be applied for the calculation of the discharge of water over the temporary weir is due to Mr. J. B. Wood, C.E., and is sufficiently accurate without being complicated:—

$$Q = 3 \cdot 33 \, (l - 0 \cdot 2 \, H) \, H^{\frac{3}{2}}.$$

Q = Quantity of water in cubic feet per second.

l = Length of weir in feet.

H = Depth of water in feet passing over the weir—

that is to say, the difference of level between the sill of
the weir and the water in the pool behind the weir.

In cases where the water approaches the weir at an
appreciable velocity the following corrected formula
is used :—

$$Q = 3\cdot33 \ (l - 0\cdot2 \ H_1) \ H_1^{\frac{3}{2}} ;$$

in which $H_1 = H + h$,

h being $\dfrac{\sqrt{2}}{64\cdot4}$

$v =$ the velocity of the approaching water in feet per
second.

Should, however, the discharge of a river or stream
be so large that a temporary dam with notch board
becomes impracticable, another method must be employed which will give rough but fairly reliable data.

A site is selected where there is a somewhat even
and regular flow of the water, and three or four careful
cross-sections are taken, being marked off on the bank
at equal distance the one from the other. Each section
would give a figure somewhat like the following, and

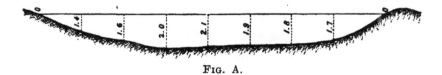

Fɪɢ. A.

the average depth would therefore be

$$\frac{0 + 1\cdot4 + 1\cdot6 + 2\cdot0 + 2\cdot1 + 1\cdot9 + 1\cdot8 + 1\cdot7 + 0}{9} = 1\cdot39$$

Should the width be sixteen feet, the section in
square feet $= 1\cdot39 \times 16 = 22\cdot24.$

The average of the three or four sections taken will therefore be the average square feet of water moving between the sections.

What is now required to ascertain the discharge of the river is the velocity of the water; to obtain this, a well arranged float that can be distinctly seen, and at least three-fourths submerged, is placed in the centre of the river some little distance above the line of the first section. On noting the time of passing the first and last sections we might have, knowing the exact distances travelled, a surface velocity, for instance, of 5·7 feet per second. A number of observations are taken, and the mean of them is the surface velocity.

The mean velocity of a river, however, is less than the surface velocity, and for rough estimations may be taken as three-fourths of the mean of the observed velocities; we should have, therefore, $5·7 \times \frac{3}{4} = 4·27$ feet per second. $22·24 \times 4·27$ is consequently the discharge, which in this case equals 94·96 cubic feet per second.

This may be taken as the minimum discharge, as four-fifths of the surface velocity is a nearer approximation to take for the mean velocity of a stream.

The following table may be considered as giving the maximum mean velocity, corresponding to the surface velocity :—

TABLE OF VELOCITIES OF STREAMS.

Surface Velocity.	Mean Velocity.	Surface Velocity.	Mean Velocity.	Surface Velocity.	Mean Velocity.
120	98·00	212·5	182·40	310	273·1
122·5	100·25	215	184·75	315	277·8
125	102·50	217·5	187·05	320	282·5
127·5	104·75	220	189·35	325	287·2
130	107·00	222·5	191·65	330	291·9
132·5	109·25	225	193·95	335	296·6
135	111·55	227·5	196·30	340	301·2
137·5	113·80	230	198·60	345	305·9
140	116·05	232·5	200·90	350	310·6
142·5	118·30	235	203·25	355	315·3
145	120·60	237·5	205·55	360	320·1
147·5	122·85	240	207·85	365	324·8
150	125·15	242·5	210·20	370	329·5
152·5	127·40	245	212·50	375	334·2
155	129·65	247·5	214·85	380	338·9
157·5	131·95	250	217·15	385	343·6
160	134·20	252·5	219·50	390	348·3
162·5	136·50	255	221·80	395	353·0
165	138·80	257·5	224·15	400	357·8
167·5	141·05	260	226·45	405	362·5
170	143·35	262·5	228·80	410	367·2
172·5	145·65	265	231·10	415	371·9
175	147·95	267·5	233·45	420	376·7
177·5	150·20	270	235·75	425	381·4
180	152·50	272·5	238·10	430	386·1
182·5	154·80	275	240·45	435	390·8
185	157·10	277·5	242·75	440	395·6
187·5	159·40	280	245·10	445	400·3
190	161·70	282·5	247·45	450	405·1
192·5	164·00	285	249·75	500	452·5
195	166·30	287·5	252·10	550	500·0
197·5	168·60	290	254·45	600	547·7
200	170·90	292·5	256·75	650	595·5
202·5	173·20	295	259·10	700	643·3
205	175·50	297·5	261·45	750	691·2
207·5	177·80	300	263·75	800	739·2
210	180·10	305	268·40		

HAVING settled upon a site that meets in every respect the requirements demanded—which embraces extensively the utilization of every source of supply, and the adaptation of such sources to the required object under investigation—the next point to be studied is the one of locating the precise position where the dam is to be built; and herein are involved a few matters that call for special consideration. It is of course understood that an impounding reservoir cannot be constructed unless there be an impervious bed under it at such a depth as to permit the construction of the dam being practicable. It is often found that where two streams meet a contraction of the valley occurs a little lower down, and the best site for the erection of a dam is where a valley widens out into a flat area bounded by steep sides, there generally being a contraction of the valley in close proximity.

The quality, condition, and class of rock; inclination of strata, permeability, and direction; whether in closing the valley the strata will pass obliquely under the dam at right angles or parallel with it—are all matters requiring consideration. Observation will also be made as to whether any portion of the district has been recently or in past times subjected to the disturbing influence of volcanic eruption or earthquake shocks. The solidity of the rock upon which the dam is to be founded is also of great importance, whilst faults or small fissures filled with clay will inevitably lead to an immense amount of excavation. It is usual, therefore, to test the ground by bores, or, preferably, shafts sunk along the line where the proposed dam is to be built. In this way the condition of the rock, the depth of the loose subsoil, and the importance of discovering any springs of water, which cause great trouble and danger in the foundations, will be at once ascertained. Too little attention is often paid to these details, with the result that as much masonry has to be placed below the surface of the ground as there is in the superstructure, with a corresponding enormous increase upon the original estimate of the work. The author has known a case where, the test shafts not being sunk deep enough, the result was that the estimated cost of the works was exceeded by twenty-three per cent.

The fall of the watercourse in which the dam is to be erected has a direct relationship to its height, a rapid course requiring a greater height of dam to give a large impounding space.

Having located the ultimate position of the dam, we can next proceed, by the aid of carefully surveyed contour lines, which have been previously set out on the ground at every three or four feet height, to calculate at various heights the capacity obtained, the required capacity indicating the necessary height of the dam.

At a point near where the various watercourses upon which we depend for supply touch the water when the reservoir is full, wells should be built of dry stone or otherwise for the purpose of retaining the silt which will be brought down by heavy rainfalls. In this way a great quantity of detritus will be checked from being swept into and depositing on the bottom of the reservoir, and so reducing its capacity. Each silt deposit can be conveniently cleaned when full, probably during dry weather, at small expense. As the bulk of the silt will be brought down during heavy storms, allowance may be made for passing off the water by a by-wash when the reservoir is full; this also relieves any undue strain upon the wall from a sudden rise of the surface of the water, and further insures that the overflow provided is under all circumstances ample.

All vegetation within the reservoir area should be collected, removed, or burnt, as the water is liable to suffer for a few years from the decaying vegetable matter. The whole catchment area must in fact be very carefully preserved, and no farming operations permitted within it, or otherwise at certain seasons of the year impurities will be found in the water which may render it dangerous for drinking purposes. A variation in the purity of the water from the famous Loch Katrine being observed, it was found on investigation to be caused by a farmer within the catchment area having, just before a very heavy fall of rain, used manure for his ground somewhat extensively.

CALCULATIONS OF STABILITY.

FOR a masonry dam to be perfectly stable the following conditions must be complied with :—

1. No part of the masonry must be in tension.
2. No part of the structure must be under more than a certain pressure from the superincumbent weight.
3. It must by friction alone resist any tendency to slide on its base.
4. It must by its own weight alone be able to resist all tendency to overturn by water pressure or the pressure of uneven winds, &c.
5. There must be such a width given to the top as to counteract the effects of expansion and contraction.

Professor Rankine points out that, theoretically, if the limits of pressure fall outside the centre third of the profile the structure is exposed to tension. The line of pressure—called also by some engineers the line of resistance—should therefore fall within the centre third, for if the requirement as to tension be fulfilled we have Conditions 2 to 4 complied with. The line of pressure is a line intersecting each joint of a structure at the point of application of the resultant of all the forces acting on that joint.

To calculate the best section that fulfils the above requirements, without any practical important excess in expenditure of material beyond what is necessary, is a very complicated problem, but with care and trouble it can be greatly simplified. The system that has been generally adopted is to make a number of trial profiles, and to adopt the one that gives the required lines. Mr. W. B. Coventry remarks, in his memoir on the " Design and Stability of Masonry Dams : "—" Owing to the indeterminate nature of the problem, it seems impossible to construct a general formula for calculating the dimensions of a dam, and the method usually followed consists in assuming an approximate profile, and then testing its stability by a graphic resolution of forces. If found defective, the profile is altered, and the graphic process repeated until a sufficiently exact result is obtained."

A correct profile may undoubtedly be found by making a number of trials, but it is extremely laborious, and may involve several days, and perhaps weeks, of unsatisfactory work, although it may be impossible to determine at once the proper section of minimum area which incorporates the required conditions; yet an approximate profile may be obtained by the application of a very simple formula giving polygonal outlines of inelegant form, and in every respect sufficiently accurate and practical to bear upon it the ultimate design. Very little attention is generally paid to the appearance of a dam, its calculated form being considered sufficient; with the result that a large expenditure may produce a structure of ungainly form, fulfilling the purpose for which it was intended, no doubt, but doing little credit to the engineer as designer. There is less intelligence required to develop on paper a structure that adheres to theoretical requirements, but the combination of security and beauty of form generally involves a higher capacity; hence our feelings are continually being outraged by the ineffective and inelegant structures around us.

The ultimate profile should adhere undoubtedly to the required theoretical profile, but there are a few things that the formula does not and cannot take into account—which, when allowed for, can be incorporated in the design of outline.

The top width of a dam is a matter of judgment, and it is this width, which cannot be calculated theoretically, that will assist the designer, if taken advantage of, to produce a pleasing and graceful form. Condition No. 5 is of great importance, and involves the following points.

Expansion and contraction are proportionate to the length, the width increasing or checking that effect according as it is wide or narrow.

The top, if vertical for eight to ten feet at the back, causes it to appear when built as if leaning over, thereby destroying when executed what may be a very pleasing effect in the original drawings.

The great heat of tropical and semi-tropical climes during the day has a very decided expanding influence upon the masonry, the author being acquainted with one masonry dam that is cracked in two places from the top downwards for a depth of eight to twelve feet, and passes a stream of water during the winter, while in the summer months it is perfectly water-tight; the top width, however, is only six feet six inches, instead of being eight feet three inches, in which case these cracks would not have taken place.

High masonry dams are usually required to close deep gorges and valleys, having consequently less length of top to depth than low masonry dams, which may impound the same quantity of water, but, being in flat and low-lying country, will be of excessive length to depth; by taking $\sqrt{H} + 2$ feet (where H equals height) for width of top, a wide top in proportion to height is obtained for low dams and a gradually decreasing width to height in high dams.

It is necessary, then, that a theoretical profile should embrace and allow for the above in all cases, which the following very simple formula does; and, that the width at quarter height from the top shall be dependent upon beauty of outline rather than strict mathematical rule. By a slight modification of Sir Guilford Molesworth's formula we have one which is applicable to both high and low dams; whilst the method employed in obtaining the offset to the inner face is very simple, and results in a very close approximation to Rankine's theoretical profile, as well as the practical or ultimate profile obtained for the inner face of the Quaker Bridge Dam, built for the supply of water for New York, which has a maximum height of upwards of 250 feet (see Fig. D, *post*, p. 41).

The following formula, therefore, may be adopted

in ascertaining any theoretical profile, and will meet rigidly the requirements of stability, which can be afterwards ascertained by the graphic system, and by calculation as a check if desired :—

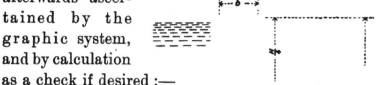

$$b = \sqrt{H+2}\text{ feet}$$

$$a = \frac{H}{4} \times 0.72$$

$$y = \sqrt{\left(\frac{0.05x^3}{\lambda + (0.03x)}\right)}$$

$$y = 0.6x + \text{ as a minimum}$$

$$z = \frac{f}{25} \text{ for 100 feet depth,}$$

$$\text{or } \left(\frac{0.09x}{\lambda}\right)^3$$

approximately for all depths.

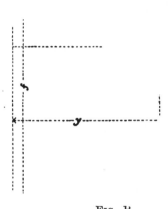

Fɪɢ. B.

Where (Fig. B) H = Total height of dam in feet.

λ = A constant of seven to be used for all depths.

x = Depth below the surface of water in feet.

y = Offset in feet to the outer face of the dam from a vertical line corresponding to the inner face at the top.

z = Offset to the inner face from the same vertical line.

The above formula will give for a dam of 100 feet in height the outline shown on the diagram of forces accompanying this work,* the dotted lines giving Molesworth's outline, the width of the top of the dam b being $0\cdot4y$ at $\frac{1}{4}$ H and the offset $z = \left(\dfrac{0\cdot09x}{\lambda}\right)^{\frac{1}{4}}$, λ being limit of pressure allowed on the masonry in tons per square foot, which is taken as seven in this case. The symbol λ, being equal to the limit of pressure allowed on the masonry, can be varied ; any less pressure, such as five or six tons, being adopted, gives a greater width to the dam ; above seven a decreased width. These ciphers are, however, empirical, and have no relation to the pressure per square foot obtained by the combined water and masonry upon the outer portion of the dam. A constant of seven has therefore been adopted in the place of any varying limit of pressure per square foot.

Having by this simple method drawn out our theoretical profile by dividing it into four sections, we can proceed to design upon this outline the ultimate form which the wall shall assume, being guided by the width at $\dfrac{H}{4}$, this width being, by the method adopted, obtained, and dependent upon an outline that combines rigidity (at the very point that is most exposed to expansion and contraction from the sun's rays) and elegance of form. This is very simply obtained by applying a curve that will start tangentially with any point at the outer face at about half the total height of the dam, passing through the point given for the quarter height and meeting the top edge ; from the half height to the base another curve will be drawn in that, closer, with the calculated widths at the half height, three-quarter height,

* For diagram, see loose sheet in pocket of cover of this book.

and the base. These two curves, it will be noticed, encroach somewhat upon the trapezium given by the formula on each section; but if the total height be divided into more than four planes the curved lines will approximate more closely to the polygonal lines of the outer face. We have, therefore, after completing this operation, an ultimate or practical profile, upon which may be based the diagram of forces, in order to ascertain the lines of resistance, with reservoir full and empty.

In order to proceed, it is necessary to settle what weight per cubic foot of masonry is to be adopted, as the calculation will be based upon a section of the dam the length of which is unity, or, in other words, a section the thickness of which is one foot. The proportions of stone to mortar in a well built dam will be two-thirds stone and one-third mortar; the stone employed should not have a less specific gravity than 2·5; the cubic metre will therefore weigh 2,500 kilogrammes. A cement mortar composed of one to three of sand—the sand being washed after passing through a one-eighth inch square hole sieve, and the weight of the cement being equal to ninety pounds per striked cubic foot—will have a specific gravity of 2·02 when fresh, but the immediate evaporation which is produced by chemical combination, and the drying which takes place afterwards, result in giving 1·99.

The density of the masonry can therefore be ascertained as follows :—

	Kilos. per c. metre.	Lbs. per c. foot.
⅔ of stone at 2,500 kilogrammes per cubic metre =	1,666	104·04
⅓ of mortar at 1,990 kilogrammes per cubic metre =	663	41·40
Total weight	2,329	145·44

The density can therefore be taken at 145lbs. per cubic foot, which will ensure, if adhered to in construction, a perfectly sound monolith; with a specially sound and heavy stone, and increasing the proportion of cement, the density would rise to 150lbs. per cubic foot, or even 155lbs. For high masonry dams of over 100 feet no more material should be used than will result in a less density than 145lbs. per cubic foot; for dams of fifty feet and less a poorer class of stone may be employed, and the masonry might have a density of not more than 130lbs. per cubic foot and be perfectly safe.

The higher the density adopted the safer the work will be, as it necessitates good material.

The next point of importance that requires consideration is the limit of pressure on the masonry at the base of the wall. It must be noticed that in buildings of this kind the mortar plays a considerable part, and on its resistance the work depends. The crushing force on an eminently hydraulic mortar composed of one of cement to three of sand, when placed under the condition in which it is found in masonry— being confined, that is to say, and subjected to lateral pressure—will be not less than 130 tons per square foot after aging. Taking one-tenth of the crushing force as the limit of pressure, we have thirteen tons per square foot. By adhering to the rule that the lines of resistance shall fall within the middle third, with the density of the masonry taken as 150lbs. per cubic foot, the pressure on the outer base of the wall will not be more than 6·35 tons per square foot at 100 feet depth, 8·11 tons at 150 feet, and 10·47 tons at

200 feet. There is no chance, therefore, of the combined masonry and water pressures exceeding this limit, whilst for dams of 200 feet depth a stronger mortar would probably be used than one to three, and thereby resist a greater crushing force.

Having determined the density, and ascertained the limit of pressure to be applied at the base, and drawn out the profile according to the amended formula, the next step is to calculate the areas, weight, and centre of gravity of the four sections into which the profile is divided, and ascertain the direction and intensity of the resultant on each plane respectively.

The determinations of the centres of gravity and pressures, &c., are adopted with modifications from the method employed by Sir Guilford Molesworth in his notes on " High Masonry Dams," the whole of which can be readily understood by the aid of the diagram appended.

To determine the centre of gravity of the masonry in each section separately :—

Let A, B, C, D (Fig. 2, Diagram) be any section (in this case the section overlaying the plane d, Fig. 1, Diagram).

Draw the diagonals $A D, C B$; S being the point of their intersection. Make $D E = A S$, join $E C$, and bisect the lines $C B$ and $E C$ at F and H respectively ; join $F E$ and $H B$; then the intersection of the lines $F E$ and $H B$ at G gives the position of the centre of gravity of the masonry above (marked also by small circles at end of arrow in each section, Fig. 1, Diagram).

To determine the position of the centre of gravity of the

masonry and of the vertical pressure of the water for each section separately :—

In any convenient position make a diagram for the polygon of forces (Fig. 3, Diagram) as follows :—

On a vertical line lay off with any convenient scale ; $JK =$ the vertical water pressure on the face AC, and with the same scale lay off $KL =$ weight of the masonry in the section A, B, C, D. Take any convenient point M (the position of M is immaterial) ; join LM, KM, and JM ; then these lines give the direction of the lines for the polygon of forces (Fig. 4), which is formed as follows :—

Bisect AC in T and draw vertical lines through T and G. Then from any convenient point, N, in the vertical line which passes through T draw the line NR parallel to MK ; then from N and R draw the lines NQ and RQ parallel to MJ and ML respectively. The point Q at the intersection of the lines NQ and RQ gives the position of a vertical line which will pass through the mean centre of gravity of the masonry and water pressure.

To determine the centre of gravity of the total pressure overlaying each plane :—

(a) Reservoir full.—The centre of gravity of each section having been found, as described above, form a diagram for the polygon of forces as before, laying off $ef = Pa$; $eg = (P + \Pi v) b$; $eh = (P + \Pi v) c$; $ej = (P + \Pi v) d$ (see Fig. 5, Diagram). Take any convenient point k, and join ek, fk, gk, hk, and jk. Then form a polygon of forces (see Fig. 6, Diagram) as follows :—

From any convenient point l in the vertical line that passes through the centre of gravity of the weights of masonry overlaying the plane a, draw $l\,q$ parallel to $e\,k$, $l\,m$ parallel to $k\,f$ until it intersects at m the vertical line which passes through the centre of gravity of the weight of masonry and vertical water pressure of that section contained between the planes a and b; then through m draw $m\,n$ parallel to $k\,g$ until it intersects at n the vertical line which passes through the centre of gravity of the weight of masonry and vertical water pressure of that section of the dam that lies between the planes b and c; then through n draw $n\,p$ parallel to $k\,h$ until it intersects at p the vertical line which passes through the centre of gravity of the weight of masonry and of vertical water pressure of the section that lies between the planes c and d, and from p draw a line parallel to $k\,j$ until it intersects the line $l\,q$. Then the points of intersection of these lines at s, r, and q give the position of vertical lines which will pass through the centre of gravity of the loads overlaying the planes b, c, and d respectively, the centre of gravity of the weight overlaying the plane a having previously been determined.

(*b*) Reservoir empty.—The centres of gravity for the empty reservoir are found in the same manner as above, laying off in the diagram for the polygon of forces P in every case instead of $P + \Pi v$. The diagram and the polygon of forces in the case of an empty reservoir are shown in dotted lines when they differ from those of the full reservoir.

To determine the resultants of pressure for each plane, the reservoir being full :—

Find the centre of horizontal pressure of water for each plane, equal two-thirds of the depth of the plane below the surface of the water; then from each point in the polygon of forces l, s, r, and q draw vertical lines to the intersection of the horizontal line of the centres of pressure respectively (these intersections are shown by the centres of circles in Fig. 1, Diagram); then from these points of intersection lay off with any convenient scale a vertical distance $= P + \Pi v$, and from the vertical distance so laid off lay off a horizontal distance $= \Pi h$; then a line from the intersection of the centres of gravity and horizontal pressure of water to the point given by this horizontal distance will represent the direction of the resultant of the weight of the masonry and of the water pressure, and where this line intersects the plane will be the centre of pressure of the resultant on the plane in question.

To determine the centre of pressure for each plane, the reservoir being empty :—

Vertical lines drawn from points v, w, y, l in the polygon of forces for the empty reservoir to the planes d, c, b, and a respectively give points in those planes for the centre of pressure.

It is obvious that the centre of gravity obtained by the above method is applicable only when each section is of trapezoidal form, as in the case of the four sections in the Diagram, Fig. 1, which is perfectly accurate for all practical purposes; should, however, it be thought desirable to adhere strictly to the final form—that is to say, the curved form on the down stream face—the following simple method may be adopted : Transfer the final and adopted profile to a piece of pasteboard, which must be carefully cut to the outline, suspend it from one of the corners of its base; and drop a vertical from the point of suspension, which line must be marked upon the profile, and a similar operation performed from the opposite side of the base ; the point where these two vertical lines intersect is the grand centre of gravity of the masonry. To obtain the same for each section the pasteboard profile must be cut along the lines of the four planes of the sections into which the whole is divided, and by so doing separate the one from the other; repeat the same operations on each section separately, as was done for the whole profile ; this will give the centre of gravity of each section.

For the purpose of checking the lines obtained by the application of the graphic system, it might be found convenient and satisfactory to derive the same result mathematically. The whole of the calculations necessary are therefore given in the following pages, and will explain themselves.

The profile, as before, is divided into four sections, and the following description corresponds to the symbols used :—

H = Height of dam.

λ = A constant of 7.

l = The length of the horizontal plane or joint in question.

P = The resultant of the weight of the overlying masonry.

Πv = The vertical component of the pressure of the overlying water.

Πh = The horizontal component of the same.

x = Depth of the imaginary horizontal plane below the surface of the water.

y = The ordinate from the vertical line to the down stream face (y_1 and y_2 being the down stream ordinates on the top and bottom of any section respectively).

z = The ordinate to the up stream face (z_1 and z_2 being the up stream ordinates on the top and bottom of any section respectively).

w = Vertical component of water pressure on each section of the dam.

W = Vertical component of pressure of masonry for each section of the dam.

g = The distance of the centre of gravity of all the masonry overlying any plane measured from the vertical line towards the down stream face, the reservoir being empty.

g^1 = The distance of the centre of gravity of all the masonry and the whole of the vertical component of water pressure overlying any plane measured from the vertical line towards the down stream face, reservoir full.

g_y = Distance of the centre of gravity of that portion of each section which lies down stream of the vertical line, measured from that line towards the down stream face.

g_z = Distance of the centre of gravity of the up stream portion of each section, measured in the up stream direction.

gm = Distance of the centre of gravity of the whole of the masonry in each section measured from the vertical line towards the down stream face.

gmw = Distance of the mean centre of gravity of the masonry and of the vertical water pressure of each section measured from the vertical line towards the down stream face.

m = Moments of each section ; or weight due to each section separately multiplied by gm or gmw respectively for empty or full reservoir.

M = Sum of the total moments overlaying each plane, reservoir empty.

M^1 = Sum of the total moments overlaying each plane, reservoir full.

c = Depth of the centre of the horizontal pressure of the water below its surface.

u = The distance of the up stream face measured on an imaginary horizontal plane from the foot of the resultant of the weight of the overlaying masonry, the reservoir being empty.

u^1 = The distance of the down stream face measured on an imaginary horizontal plane from the foot of the resultant of the pressure of the water and of the weight of the overlaying masonry.

p = Resultant pressures of masonry and water on bearing surface of each plane; the bearing surface being from a vertical line passing through the centre of gravity of the weight of masonry and vertical water pressure of each section to down stream face.

DETAILED CALCULATION OF MASONRY DAM OF 100 FEET HEIGHT.

DENSITY OF MASONRY 145 LBS. PER CUBIC FOOT = (0·06473 TONS PER C. FT.).

No.	Description	Symbol	a	b	c	d	Notes
					Imaginary Planes.		
[1]	$x = $	x	25	50	75	100	Feet.
	$y = \sqrt{\dfrac{0{\cdot}05\,x^3}{\lambda + (0{\cdot}03\,x)}} = $		10·04	27·12	47·75	70·71	"
	$y = $ as a minimum $0{\cdot}6\,x$		15	30	45	60	"
	Width at top $= \sqrt{/H} + 2 = 12$	b					
	$y = $ at $\frac{1}{4} H = \dfrac{H}{4} \times 0{\cdot}72 = $	a	18				"
[2]	y as adopted $= $	y	18	30	47·75	70·71	"
[3]	$z = $ Batter of 1 in 25 from $\dfrac{H}{4}$ downwards $= $	z	0	1	2	3	"
[4]	Length of point $= y + z = $	l	18	31	49·75	73·71	"
[5]	Mean width of each section $= $		15	24·5	40·375	61·73	"
[6]	Area of each section $= [5] \times h = $		375	612·5	1,009·875	1,543·25	Square feet, $h = 25$ ft.
[7]	Weight of each section 1 ft. wide $= [6] \times 0{\cdot}06473$ tons $= $	W	24·274	39·647	65·387	99·895	Tons.
[8]	Weight of masonry above each plane $= $	P	24·274	63·921	129·258	229·153	"
[9]	Vertical pressure of water on each section $= $	w	0				Weight of 1 c. ft. water $= 0{\cdot}02787$ tons.
	$z_2 - z_1 \left\{ x_2 - \dfrac{x_2 - x_1}{2} \right\} \times 0{\cdot}02787$ tons $= $	"					
	$1 - 0 \left\{ 50 - \dfrac{50-25}{2} \right\} \times 0{\cdot}02787$ " $= $	"		1·05			Tons.
	$2 - 1 \left\{ 75 - \dfrac{75-50}{2} \right\} \times 0{\cdot}02787$ " $= $	"			1·74		"
	$3 - 2 \left\{ 100 - \dfrac{160-75}{2} \right\} \times 0{\cdot}02787$ " $= $	"				2·44	"

No.	Description	Symbol					Notes
[10]	Vertical component of water pressure on each plane = ...	Πv	0	1·05	2·79	5·23	„
[11]	Vertical press of masonry and water on each plane [8] + [10] = ...	$\dfrac{P+\Pi v}{\Pi h}$	24·274	64·97	132·05	234·38	„ 0·013934 = weight of 1 c. ft. of water in tons divided by 2. Tons.
[12]	Horizontal water pressure = $0·013934\,x^2$ = ...						
	$0·013934 \times 25^2$ = ...	„	8·71				„
	$0·013934 \times 50^2$ = ...	„		34·84			„
	$0·013934 \times 75^2$ = ...	„			78·38		„
	$0·013934 \times 100^2$ = ...	„				139·84	„
	Depth of centre of pressure = $\frac{2}{3}x$ = ...	c	16·67	33·33	50	66·67	Feet.
[13]	Coefficient of sliding forces = $\dfrac{[12]}{[11]}$ = ...	$\dfrac{\Pi h}{P+\Pi v}$					
	$\dfrac{8·71}{24·27}$ = ...		0·359				⎫
	$\dfrac{34·84}{64·97}$ = ...			0·596			⎬ Limit = 0·76, which is the tangent of the angle of repose of masonry.
	$\dfrac{78·38}{132·05}$ = ...				0·594		⎪
	$\dfrac{199·84}{234·38}$ = ...					0·595	⎭

Details of the determination of the Centres of Gravity, &c., by calculation.

Formula	Symbol					Notes
$\frac{1}{3}\left(y_1+y_2 - \dfrac{y_1 y_2}{y_1+y_2}\right)$ =	g_v					
$\frac{1}{3}\left(12+18 - \dfrac{12\times18}{12+18}\right)$ = ...	„	7·60				Feet.
$\frac{1}{3}\left(18+30 - \dfrac{18\times30}{18+30}\right)$ = ...	„		12·25			„
$\frac{1}{3}\left(30+47·75 - \dfrac{30\times47·75}{30+47·75}\right)$ = ...	„			19·78		„
$\frac{1}{3}\left(47·75+70·71 - \dfrac{47·75\times70·71}{47·75+70·71}\right)$ =	„				29·99	„

CALCULATION OF MASONRY DAM OF 100 FEET HEIGHT.

OF MASONRY 115 LBS. PER CUBIC FOOT = (0·0643 TONS PER C. FT.).

Description	Symbol	Imaginary Planes				Notes.
		a.	*b.*	*c.*	*d.*	
	x	25	50	75	100	Feet.
		10·04	27·12	47·75	70·71	,,
	b	15	30	45	60	,,
	a	18				,,
	y	18	30	47·75	70·71	,,
	z	0	1	2	3	,,
	$/$	18	31	49·75	73·71	,,
		15	24·5	40·375	61·73	,,
			0·25	1,009·375	1,543·25	Square feet, h = 25 ft.
				65·337	99·895	Tons.
					229·153	,,

Reservoir

$(W+w)\ gmu =$ ....

$(24\cdot27 +0)\ 7\cdot60 =$ ....

$(89\cdot65 +1\cdot05)\ 11\cdot67 =$ ....

$(65\cdot84 +1\cdot74)\ 18\cdot49 =$ ....

$(99\cdot90 +2\cdot44)\ 27\cdot98 =$ ....

Reservoir full $= \Sigma m^1 =$ ....

14 $\dfrac{M}{P} =$ Reservoir empty $=$ ....

$\dfrac{184\cdot45}{24\cdot27} =$ .... $7\cdot60$

$\dfrac{659\cdot85}{68\cdot92} =$ .... $14\cdot72$

$\dfrac{1902\cdot62}{129\cdot26} =$ .... $20\cdot82$

$\dfrac{4771\cdot75}{229\cdot15} =$

15 $\dfrac{M^1}{P+\Pi w} =$ Reservoir full $=$ ....

$\dfrac{184\cdot45}{24\cdot27} =$ ....

$\dfrac{659\cdot42}{64\cdot07} =$ .... $14\cdot39$

$\dfrac{1899\cdot73}{132\cdot05} =$ .... $20\cdot82$

$\dfrac{4768\cdot20}{284\cdot98} =$

DETAILED CALCULATION OF MASONRY DAM OF 100 FEET HEIGHT—(continued).

No.	Description.	Symbol.	Imaginary Planes.				Notes.
			a.	b.	c.	d.	
[16]	$z + g = $	u	7·60	11·32	16·72	23·82	Feet. Reservoir empty.
[17]	$y - \left\{ g^1 + (x - c)\dfrac{\Pi h}{P + \Pi v} \right\} = $ ...... $18 - \{7·60 + (25 - 6·67)\,0·359\} = $ ...... $30 - \{10·15 + (50 - 33·33)\,·566\} = $ ...... $47·75 - \{14·89 + (75 - 50)\,·894\} = $ ...... $70·71 - \{20·32 + (100 - 66·7)\,0·595\} = $	u^1	7·41	10·91	18·51	30·56	Reservoir full.
[18]	$\dfrac{\sqrt{\{(P + \Pi v)^2 + \Pi h^2\}}}{l - (z + g)} = $ ...... $\dfrac{\sqrt{\{(24·274)^2 + 8·71^2\}}}{18 - (0 + 7·60)} = \dfrac{25·79}{10·40} = $ ...... $\dfrac{\sqrt{\{(64·97)^2 + 34·84^2\}}}{31 - (1 + 10·15)} = \dfrac{73·72}{19·85} = $ ...... $\dfrac{\sqrt{\{(132·05)^2 + 78·88^2\}}}{49·75 - (2 + 14·89)} = \dfrac{153·56}{33·36} = $ ...... $\dfrac{\sqrt{\{(234·38)^2 + 189·34^2\}}}{73·71 - (3 + 20·32)} = \dfrac{272·67}{50·39} = $	p	2·48	3·71	4·60	5·41	Tons. Maximum pressure per square foot.

DETAILED CALCULATION OF MASONRY DAM OF 200 FEET HEIGHT.

DENSITY OF MASONRY 150 LBS. PER CUBIC FOOT (0·6696 TONS PER C. FT.).

No.	Description	Symbol	Imaginary Planes.				Notes.
			a.	b.	c.	d.	
[1]	$x =$		50	100	150	200	Feet.
	$y = \sqrt{\dfrac{0·05 x^3}{\lambda + (0·08 x)}} =$		27·11	70·71	121·14	175·41	,,
	$y =$ as a minimum $0·6 x =$		30	60	90	120	,,
	Width at top $= \sqrt{H} + 2 = 16·14 =$	b					,,
	$y =$ at $\frac{1}{4}$ $H = \dfrac{H}{4} \times 0·72 =$	a	36				,,
[2]	y as adopted $=$	y	36	70·71	121·14	175·41	,,
[3]	$z =$ from $\frac{H}{4}$ to 100 ft. depth a batter of 1 in 25 $=$ ...	z	0	2	7	17	,,
	,, 100 ft. to 150 ft. ,, ,, 1 in 10 $=$ ...	,,					,,
	,, 150 ft. to 200 ft. ,, ,, 1 in 5 $=$ ...	,,					,,
[4]	Length of joint $= y + z =$	,,	36	72·71	128·14	192·41	,,
[5]	Mean width of each section $=$	,,	26·07	54·355	100·425	160·275	,,
[6]	Area of each section $= [5] \times h =$	h	1303·50	2717·75	5021·25	8013·75	Square feet $h = 50$ ft.
[7]	Weight of each section 1 ft. wide $= [6] \times 0·06696$ tons $=$	W	87·282	181·981	336·223	536·601	Tons.
[8]	Weight of masonry above each plane $=$	P	87·282	269·263	605·486	1142·087	,,
[9]	Vertical pressure of water on each section $=$ $z_2 - z_1 \left(z_2 - \dfrac{z_2 - z_1}{2} \right) \times 0·02787$ tons $=$	w	0	4·18	17·42	48·77	,, weight of 1 c. ft. of water $=$ 0·02787 tons.
[10]	Vertical component of water pressure on each plane	Πv	0	4·18	21·60	70·87	,,
[11]	Vertical press of masonry and water on each plane $= [8] + [10] =$	$P + \Pi v$	87·282	278·448	627·086	1212·457	,,
[12]	Horizontal water pressure $= 0·013994\, x^2 =$	Πh	34·84	139·34	313·52	557·96	,, weight of 1 c. ft. of water in tons divided by 2 $=$ 0·013994.

DETAILED CALCULATION OF MASONRY DAM OF 200 FEET HEIGHT—(*continued*).

No	Description	Symbol	Imaginary Planes.				Notes.
			$a.$	$b.$	$c.$	$d.$	
[13]	Depth of centre of pressure $=\frac{2}{3}x=$	c	33·33	66·67	100	188·33	Feet.
	Coefficient of sliding forces $=\dfrac{[12]}{[11]}=$	$\dfrac{\Pi h}{P+\Pi v}$	0·399	0·510	0·500	0·460	Limit $=0.76$.
	Details of the determination of the Centres of Gravity, &c., by calculation.						
	$\frac{1}{3}\left(y_1+y_3-\dfrac{y_1 y_3}{y_1+y_3}\right)=$	g_y	13·67	27·62	49·07	74·97	Feet.
	$\frac{1}{3}\left(z_1+z_3-\dfrac{z_1 z_3}{z_1+z_3}\right)=$	g_z	0	0·67	2·48	6·35	,, vertical face.
	$g_y-\dfrac{z_1+z_3\,(g_y+g_z)}{w\,(z_3-z_1+g_m)}=$	g_m	13·67	27·10	46·76	68·88	,, ,, ,,
	$g_m-\dfrac{\dfrac{W}{2}}{W+w}=$	g_{mw}	13·67	26·47	44·23	62·14	,, ,, ,,
	$W\times g_m=$	m	1193·12	4931·66	15721·65	36961·01	Reservoir empty.
	Reservoir empty $=\Sigma m=$	M	1193·12	6124·78	21846·43	58807·44	Sum of moments.
	$(M+w)\,g_{mw}=\Sigma m^1=$	m^1	1193·12	4927·66	15641·50	36374·89	Reservoir full.
	Reservoir full $=\Sigma m^1=$	M^1	1193·12	6120·78	21762·28	58137·17	Sum of moments.
[14]	$\dfrac{M}{P}=$ Reservoir empty $=$	g	13·67	22·75	36·08	51·49	Feet.
[15]	$\dfrac{M^1}{P+\Pi v}=$ Reservoir full $=$	g^1	18·67	22·38	34·70	47·95	,,
[16]	$z+g=$	u	13·67	24·75	43·08	68·49	,, reservoir empty.
[17]	$y-\left\{g^1+\left(x-c\,\dfrac{\Pi h}{P+\Pi v}\right)\right\}=$	u^1	15·68	31·88	61·44	96·79	,, reservoir full.
[18]	$\dfrac{\sqrt{\{(P+\Pi v)^2+\Pi h^2\}}}{l-(z+g^1)}=$	p	4·21	6·35	8·11	10·47	Tons, maximum press. per square foot.

The level of the water for the purpose of the calculations is the same as that of the top of the wall; there is therefore a small margin of safety by this practice. As the sill of the waste weir would be below the top level of the wall, the water of the reservoir would never rise to this level unless some stupendous and incalculable rainfall take place within the drainage area of the dam. It is safest, however, to allow for all contingencies, and it is consequently advisable to follow this practice in all cases.

In the Diagram (Fig. 1) it will be noticed that the line of resistance, reservoir empty, falls outside the middle third near to the base. This is of no practical importance in this case, firstly, by reason of its being only nine inches outside; secondly, at this depth the reservoir will probably always contain water, the sluice valve being somewhat above the lowest point in the dam, whilst the discharge pipe will always be considerably above that level. There will consequently be water in the lower basin of the reservoir, and the dam cannot be considered at this point as being empty. And, thirdly, a very small increase in the density of the masonry will tend to draw together the lines of resistance for both reservoir full and empty.

Comparison of Professor Rankine's Profile with the Practical Profile.

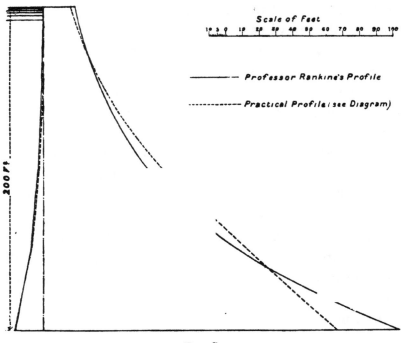

Fig. C.

CURVED masonry dams are undoubtedly of greater strength than straight dams, as it is evident that on inclosing a narrow gorge with a curved dam, convex on the up stream face, it would act as an arch abutting on the sides of the valley, and in consequence of its form adapt itself to the variations of expansion and contraction. The inner face would not be exposed to tension due to any force acting upon it, whilst the uncertain expansive force of ice which might be formed on the surface of the reservoir would be amply resisted; but the advantage which this form has over that of a straight dam depends upon the radius of the curve. A long or short radius is indefinite, and would depend somewhat upon the height of the dam; under 300 feet may, however, be considered a short radius, and over 600 feet a long one; beyond this limit any advantage from the curved form would be doubtful.

It might be generally thought that by adopting this form material could be economised by reducing the sectional area of the profile. There is a general consensus of opinion, however, that the same principle should be followed in designing a profile, whatever the plan, unless a curve of less than 300 feet be adopted; as it is found impossible to state definitely when a dam may be considered to act as an horizontal arch, the investigation resting on assumption, and being, withal, exceedingly complicated.

The board of experts—Messrs. J. P. Davis, J. J. R. Croes, and W. F. Shunk—who were appointed by the Aqueduct Commissioners to the Quaker Bridge Dam, reported upon the design of the profile as follows:—

"1. That, in designing a dam to close a deep, narrow gorge, it is safe to give a curved form in plan, and to rely upon arch action for its stability; if the radius is short, the cross-section of the dam may be produced below what is termed the gravity section, meaning thereby a cross-section or profile of such proportions that it is able, by the force of gravity alone, to resist the forces tending to overturn it, or to slide it on its base at any point.

"2. That a gravity dam, built in plan on a curve of long radius, derives no appreciable aid from arch action so long as the masonry remains intact; but that, in case of a yielding of the masonry, the curved form might prove of advantage.

"The division between what may be called a long radius and what may be called a short radius is of course indefinite, and depends somewhat upon the height of the dam. In a general way, we would speak of a radius under 300 feet as a short one, and one of over 600 feet as a long one, for a dam of the height herein contemplated.

"3. That, in a structure of the magnitude and importance of the Quaker Bridge Dam, the question of producing a pleasing architectural effect is second only to that of structural stability, and that such an effect can be better obtained by a plan curved regularly on a long radius than by a plan composed of straight lines with sharp angular deflections.

"4. That the curved form better accommodates itself to change of volume due to change of temperature.

"While chance of the rupture of the masonry of the dam by extraordinary. forces, if built on the profile herein recommended, is, in our opinion, very remote, yet it exists; and because it exists, and because the curved form is more pleasing to the eye, better satisfies the mind as to the stability of the structure, and more readily accommodates itself to changes of temperature, we think that it should be preferred in any case where it would cause no great addition to the cost.

"In comparing different locations of the dam, in order to discover the one which combined most effectively the advantages of economical construction and pleasing effect, we were confronted with the fact that—our calculations indicate that—in a dam built upon a curved plan of large radius the bottom down stream toe pressures are increased beyond those in a straight dam of the same section, in consequence of the length of the toe being less than the length of the face to which the pressure is applied. This increase of pressure is not exactly proportional to the decrease of length of toe, but it is of such magnitude that it should not be neglected in designing the section of the dam; and it involves the necessity of increasing the mass of masonry in a certain proportion to the radius of the curvature.

" *Conclusions.* — In view of the premises, and pursuant to our instructions, and believing that the dam will be more pleasing in appearance and better able to resist extraordinary forces if built on a curved plan, and bearing in mind that an excessive thrust in the direction of the curve cannot be produced until the force of gravity has been overcome, and that the profile is so proportioned that more than twice the greatest pressure exerted by any conceivable ordinary force is necessary to overcome the resistance of gravity, we recommend the adoption of the profile (Fig. D.), and of a curved plan on a radius of about 1,200 feet, as hereinbefore described, and we advise that the exact line be determined after further borings shall have established the most desirable location on the conditions prescribed.

" It should be added, in conclusion, that the form and dimensions herein recommended for adoption are prescribed on the assumption that the structure shall be well founded, and that its material and workmanship shall be of the first class in their several kinds."

PROFILE OF QUAKER BRIDGE DAM,

designed by Board of Experts.

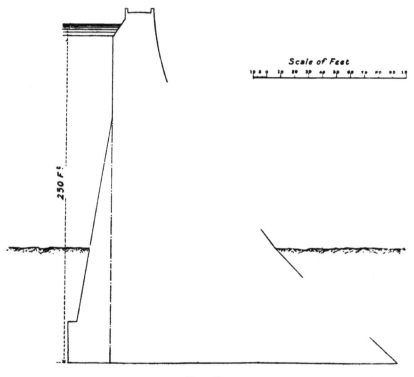

FIG. D.

WASTE WEIRS.

THE waste weir of a dam is of great importance, although not to such an extent as for earthwork dams; an ample allowance must, however, be given. To definitely fix what actual width should be allowed for safety, three points must be taken into consideration :—

1. The maximum flow of a stream is never equal to the quantity of water falling over the catchment area.

2. The reservoir itself acts as an accumulator to the extent of the rise of the water over the sill of the weir.

3. The measured height of water above the overflow level due to storms is further raised by wave action.

The capacity of the reservoir is up to the sill of the weir. Any rise above this level is therefore what the weir discharges *plus* the water accumulated by that height; and, as considerable depth of weir increases the amount of masonry along the whole top of a dam, it is usual to make the width considerable to obviate this objection, as well as the danger from choking by trees and timber that may be brought down by the flood.

The discharge of a waste weir with paved apron may be calculated from the following formula :—

$$V = \sqrt{\frac{K\,D}{S}} \qquad Q = A\,V$$

where

V = Mean velocity in feet per second.

D = Mean hydraulic depth = $\dfrac{\text{Sectional area of flow}}{\text{Wetted perimeter.}}$

S = Slope or length of channel to fall of 1.

K = 8,500.

A = Sectional area of flow.

Q = Cubic feet of water discharged per second.

A table given by **Mr. W. K. Burton, C.E.**, in his admirable work on " The Water Supply of Towns," and based on one by Fanning, will be found useful, the depths to width being safe in countries not liable to tropical rains :—

Catchment area.	Length of weir.	Depth of water over weir.
200 acres.	16 feet.	1 foot 5 inches.
400 ,,	20 ,,	1 ,, 10 ,,
1 square mile.	25 ,,	2 feet.
2 ,, miles.	32 ,,	2 ,, 5 inch.
3 ,, ,,	39 ,,	2 ,, 7 inches.
4 ,, ,,	44 ,,	2 ,, 9 ,,
6 ,, ,,	54 ,,	3 ,,
8 ,, ,,	61 ,,	3 ,, 3 ,,
10 ,, ,,	68 ,,	3 ,, 6 ,,
15 ,, ,,	83 ,,	3 ,, 9 ,,
20 ,, ,,	95 ,,	4 ,,
25 ,, ,,	105 ,,	4 ,, 2 ,,
30 ,, ,,	116 ,,	4 ,, 4 ,,
40 ,, ,,	133 ,,	4 ,, 7 ,,
50 ,, ,,	149 ,,	4 ,, 10 ,,
75 ,, ,,	183 ,,	5 ,, 3 ,,
100 ,, ,,	212 ,,	5 ,, 8 ,,

For intermediate areas the length and depths may be taken as proportionate.

The effect of wave action on a well constructed masonry dam may be ignored, but provision should be made for the free break of the waves over the top. Any protection wall along the top is objectionable, as it receives the full blow of the waves. A free top with a slope to the back to allow the water to run freely, with specially prepared drains at the base to convey the water into the valley below, is perfectly safe.

If the surface area of the water impounded be great, it may be necessary and advisable to calculate the height of wave to which the dam might be exposed, and for that purpose Stevenson's formula will be found useful. It is based on the fact that the maximum height of a wave is a function of the "fetch" of the reservoir, the fetch being the longest straight line that can be measured from any part of the dam to any part of the shore of the reservoir when full :—

$$H = 1 \cdot 5 \sqrt{D} + (2 \cdot 5 - \sqrt[4]{D}).$$

H = Height of wave in feet.
D = Fetch in miles.

It is obvious that, should the reservoir be of very great length, so as to cause high waves, the waste weir must be increased, or otherwise too large a volume of water will dash over the top, which will undermine the base and endanger the structure.

THERE are many methods adopted to supply water from reservoirs. For low dams a cast iron tower made in segments and bolted together, with inlets at various heights, answers the purpose very well; in other cases a simple bronze valve fixed into the discharge pipe, with raising and lowering gear actuated from the top of the wall, with movable cage or screen over the valve to prevent obstruction or damage, is of small cost and effective. Simplicity of construction is essential, and for high dams no better form can be employed than a round masonry tower starting from the toe of the inner face of the dam, connecting with a gallery that passes through the dam constructed for the purpose of receiving the discharge pipes, which would rise in the valve tower, where at various heights branches are attached that are controlled by exterior as well as interior valves.

This system permits of any repair being done without loss of water or danger to workmen. From want of forethought in designing the discharge arrangements for reservoirs, not infrequently divers have had to be employed to either remove obstruction or adjust the valves. The diameter of the tower is governed by the size of the discharge pipes which it is to contain, allowing for descending, or ascending and manipulation of the various heavy pieces during erection and repair.

The tower must be built of well cut and bonded stone, as it will be subjected to considerable pressure, and special care must be taken that a secure connection is made to the discharge gallery.

For estimating the diameter of pipe required for a known supply derived from different heads of water, Eytelwein's formula may be employed :—

$$W = 4 \cdot 71 \sqrt{\frac{D^5 H}{L}} \qquad\qquad D = 0 \cdot 538 \sqrt[5]{\frac{L W^2}{H}}$$

where

D = Diameter of pipe in inches.

H = Head of water in feet.

L = Length of pipe in feet.

W = Cubic feet of water discharged per minute.

In no case should the discharge pipe be at the lowest part of the valley, space being required for silt, and at this point is generally built into the wall a large scour pipe with deep and heavy flanges, to form a perfect bond with the masonry of the dam, and prevent the possibility of any percolation of water between it and the pipe. A heavy bronze sluice valve will cover the face of this pipe, actuated from the top, and an arrangement may be required to agitate the sludge around its mouth, the pressure of the water when once a blow is started being sufficient to clean the bottom of any accumulated silt. It is the practice of some engineers to leave a well in the centre of the dam, connection being made through the wall to the face at various heights, to which the discharge pipe is attached, the scour pipe being at the bottom, a gallery from the well to the outer face being utilised for the double system. This method cannot, however, be recommended, by reason of its weakening effect upon the stability of the dam.

The Alicante Dam, which was built during the years 1579 to 1594, and ascribed to Herreras, the famous architect of the Escurial Palace, is in a deep gorge, being 134½ feet high, and its length at the top only 190 feet. The plan is curvilinear, having a radius on the up stream side of the crown of 351 feet. In Mr. E. Wegmann's valuable work on " The Design and Construction of Masonry Dams," the following descriptive remarks are given of the scouring arrangements :—

" Owing to the steep declivity of the beds of most Spanish streams and to violent storms, large quantities of fine material, which has been pulverized by the action of the water, are deposited in the storage reservoirs. Unless some means were provided to remove this sediment it would soon fill these basins completely. In 1843, when the Alicante reservoir had not been cleaned for fourteen years, a bank of sediment 75 feet high at the dam had been deposited. Since then the reservoir is scoured once in four years, the maximum height of the material deposited during that time being 39 to 52 feet.

" Long experience has taught the Spaniards the best method of removing these deposits, namely, by means of scouring galleries. In the Alicante Dam such a gallery is placed in the axis of the valley, crossing the dam in a straight line from face to face. Its up stream opening is 1·8 metres wide by 2·7 metres high. The gallery has this cross-section for the first 2·7 metres of its length, and is then suddenly enlarged to a section of 3 metres width by 3·3 metres height. After this the cross-section is increased gradually, so that it is 4 metres wide by 5·85 metres high at the down stream face of the dam. By this increase in the cross-section of the gallery, which takes place in all directions, the material forced out of the reservoir by the water pressure can expand freely, and does not obstruct the channel through the dam.

" The mouth of the scouring gallery is closed simply by a timber bulkhead formed as follows: First, a vertical row of beams about one foot square is placed, their ends projecting into horizontal grooves cut into the solid masonry; the last beam which closes the row is somewhat shorter than the rest, and enters only the lower groove. After the joints between the beams have been calked a second row of similar timbers is placed directly behind the first row, they are laid horizontally, their ends being secured in vertical grooves in the sides of the gallery. Behind the second row three vertical posts are placed, each of which is firmly held by two inclined braces whose lower ends project into the floor of the gallery.

" The banks of sediment formed in the reservoir acquire considerable consistency if left undisturbed for a few years. When it is necessary to scour the reservoir it becomes thus possible to remove gradually the

timbers at the inlet of the gallery without much
danger to the workmen. The timbers of the course
next the reservoir are cut, one by one, with the
greatest precaution. Should any movement be per-
ceptible in the deposited material the men abandon
their work, which will be quickly completed by the
water pressure.

" Generally, however, the opposite to this takes place.
The sediment forms a solid bank in front of the
scouring gallery, and does not move until a hole has
been made through it from the top of the dam. The
heavy iron bar which is employed for this purpose at
the Alicante reservoir is 0·2 feet square, 59 feet long,
and weighs about 1,100 lbs. It is worked by means of
a windlass and pulleys. When a hole has been pierced
through the bank of sediment the scouring action
begins, first slowly, but soon gaining a tremendous
force. All the sediment, except that in remote parts
of the reservoir, is forced through the scouring gallery,
the noise made by this violent action being like that of
cannons. Nothing remains for the workmen to do but
to shovel the remaining sediment into the current.
Sometimes the deposit has become so hard that it must
be undermined from the scouring gallery before a hole
is pierced in it by the long bar. The total cost of
scouring the reservoir, including the loss of timbers
which are cut, amounts to only £10.

" Although the method of cleaning the reservoir seems
at first sight rather primitive, yet, on second thoughts,
it will be found to be practical. Where such deep
deposits are made gates are out of the question, as they

would have to be frequently opened to prevent their becoming useless, and would thus cause a considerable loss of water. While the scouring operation as carried on at the Alicante Dam certainly involves danger to the workmen, accidents are very rare."

HAVING calculated the requisite profile and completed the design of the dam, it becomes necessary to settle upon the form of foundations that shall be adopted, and in what way all possibility of filtration shall be prevented. However carefully this may be schemed out, modification may become necessary during the progress of the work. Should any unexpected variation in the condition of the rock present itself in opening up the foundations, all loose and weathered portions of the rock will require to be stripped off to the solid for the whole width of the dam at its base and cut into horizontal beds. There will, however, when this is done, still be small cracks, false seams, &c., left, which though probably not apparent to the eye, are still there, and possibly of such a length as to pass from the face to back of dam. For further securing the foundations against percolation a trench should be cut of not less than ten feet width at the top and three feet at the bottom, being of such a depth as to completely cut out any false seam, spring, or vein of soft material, clay, &c.; the relative position of this trench being in the centre of the base-width of dam, or a third of the width from the inner toe, and parallel to it for the whole length of the foundations. The position of this trench in high dams is of importance, as the upthrust from the pressure of water tends to counterbalance the downthrust of the masonry, which may seriously affect the stability.   In some dams this apron or valance, being excavated along the inner toe of the wall, leaves the whole foundation dry and free from any force which the water might exert upon the base when the reservoir is full.

In the lower portion or centre of the valley or gorge water will be met with, which must be concentrated and dammed back by well calked timbering or other methods, that the foundation material, whatever it may be, can be built into position free of water. Wherever a percolation in the rock is observed the trench must be excavated below that point, so that the spring may be seen issuing above the lowest part of the foundation.   There is always greater security in working in the centre of the valley than in rising on each side, as porosity is made evident by the water and moisture.   The experience gained, however, by close observation of the condition of the rock at the lowest level will guide the engineer as to the necessary depth to cut the trench in the higher portions.

The foundation rock, before building operations begin, should be well cleaned by the use of brushes made of steel wires, and washed by a stream of water from a hydrant or pump that will give force enough to clean the crevices of the rock of all sand or dirt and assist the effect of the brushes.

Blasting operations in the foundations should be

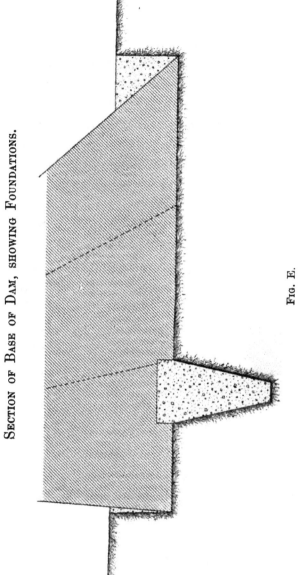

SECTION OF BASE OF DAM, SHOWING FOUNDATIONS.

FIG. E.

Scale, one-sixteenth of an inch = one foot.

conducted with great care and supervision, and the trench cut without the use of any kind of blasting material. The inner portion of the basin, should it appear necessary, may be cleared of subsoil, and seams containing clay, &c., cleared out and sealed with cement mortar. This is a further security against percolation, though a tedious and expensive operation, which rarely has to be resorted to, and should not be done to save excavation under the base of the dam, as it will in any case only prevent considerable leakage, but not percolation; whereas, if the trench be sunk deep enough, all filtration will be cut off.

The Puentes Dam in Spain, which was 164 feet high, was totally destroyed eleven years after it was built from having faulty foundations. It was originally intended to found the wall entirely on rock, but in the centre of the valley a deep pocket of subsoil was encountered, and it was decided to build the wall at this point on a pile foundation. This was built very securely, and would have answered the purpose if the water had not risen above eighty-two feet, at which level it was maintained during the eleven years. In April, 1802, however, the water rose to an elevation of 154 feet above the base of the dam, and the foundations gave way. It was noticed, just before the wall burst, that on the down stream side water of a red colour was issuing in great quantities. In a short time an explosion occurred, and enveloped by an enormous body of water, the piles and timbers which formed the pile-work of the foundation and apron were forced upwards; the volume of water that escaped was so considerable that the reservoir was emptied in the course of an hour; 608 lives were lost, and property amounting to £210,000 was destroyed.

We have here demonstrated the important fact that a high masonry dam, however well proportioned, can only be safe if founded entirely on solid rock.

CONSTRUCTION.

In construction we have the choice of four classes of masonry :—

1. Cut-stone masonry (block masonry).
2. Rubble masonry.
3. Concrete.
4. Rubble or concrete with cut stone (ashlar) facings.

1. The first would seem to possess the best and most solid material for the purpose, and be of great strength. Its cost, however, is between three and four times that of rubble, and it offers only about twice the strength.

2. Rubble masonry is easily worked, lends itself safely to the treatment of unskilled labour, possesses ample strength, and is readily adaptable to any form of profile.

3. Concrete has all the advantages of rubble masonry; though considered to be too pervious, it has, however, been used successfully, and is undoubtedly the most easily worked of these three classes of masonry. Rubble masonry is, after all, only concrete with a large aggregate.

4. Rubble or concrete with cut stone facings cannot be recommended, in consequence of its liability to uneven settlement.

Rubble masonry is, undoubtedly, generally the most convenient to use, and probably the most secure against percolation. Regular horizontal courses must be avoided, the great object being to form a monolith as homogeneous as possible. The stones to be employed may be from half a cubic foot, as in the Tausa Dam, to six or seven cubic feet, and should be laid with their natural bed horizontal. Whatever strength of mortar may be used, a certain amount of leakage will always take place when the reservoir is first filled. In a well constructed dam this leakage only shows itself as a dampness on the back face, and generally disappears in the course of six months or a year.

Great care should be exercised in the selection of the stone that is to be employed in the construction. It will, however, be generally found, as before stated, that adhering to 150 lbs. per cubic foot as the density will insure a good class of stone being used. A few tests may, however, be necessary, and we may therefore note the following :—

The carbonic acid which exists in the pure atmosphere of the country decomposes any stone of which either carbonate of lime or carbonate of magnesia forms a considerable part.

The oxygen of ordinary air will also act upon a stone containing much iron.

An examination of the stone may be made by magnification. A recent fracture when seen through a magnifying glass should be bright, clean, and sharp, with the grains well cemented together. A dull earthy appearance betokens a stone likely to decay.

The amount of water a stone will absorb is also an indication as to its quality, the best stone absorbing the least quantity of water. This can easily be ascertained by immersing the stone it is required to test in water for twenty-four hours, and noting the weight both before and after immersion. By immersing small pieces of stone in a concentrated boiling solution of sulphate of soda (Glauber's salts), and then hanging them up for a few days in the air, the salt will crystallize in the pores of the stone, forcing off pieces from the corners; and should the stone have open seams or joints, it will detach large fragments. By weighing the stone used, both before and after this test, the amount of forced disintegration can be ascertained.

Carbonate of lime or carbonate of magnesia can be easily detected by dropping on the surface of the stone a very small quantity of hydrochloric acid, when an intense effervescence will be noted if there be present a large proportion of either carbonate.

In quarrying stone for building purposes the least amount of blasting is preferable, the stone becoming greatly shaken by any sudden and violent explosion, and the waste produced proportionally large. All stones intended for building purposes should be placed in the work with their natural beds at right angles to the pressure that will come upon them. This should therefore govern the mode of operation in winning the stone.

In testing the resistance to crushing of various stones care must be taken that the samples used are not too small. The result obtained will generally indicate somewhat more than the real strength of the material, in consequence of the fracture taking place by shearing on a plane inclined at a slope having one-and-a-half rise to one of base; the experiments should therefore be made on prisms whose heights are about one-and-a-half times their diameters. Basalts, primary limestone, slates, &c., give way suddenly; other stones begin to crack under a-half to two-thirds the crushing load.

It is generally laid down that the compression to which a stone should be subjected in a structure should not exceed one-tenth of the crushing weight as found by experiment. The weakest sandstones that exist will bear a compression of 120 tons per square foot. The resistance of ordinary building stone ranges from 140 to 500 tons per square foot; granite and traps rising as high as 700 or 800 tons per square foot.

It is important that the stones used on the face or back of the dam should contain no soft patches or inequalities, but be specially selected, and that the whole of the stone from the quarry used in the work should be stacked and exposed to air action for a few weeks before being employed for building.

We have now to consider the class of mortar and its composition that is to be used for cementing together the stone which is to be employed in the construction of a dam. It should be remembered that the presence of moisture in hydraulic limes and cements favours the continuance of the formation of the silicates, &c., commenced in the kiln, and that their setting action is prematurely stopped if they are allowed to dry too quickly. It is therefore of the utmost importance, especially in hot climates, that the stones to be embedded in the mortar should be well wetted, so that they will not absorb the moisture from the mortar, and also to remove the dust from their faces, which would prevent the mortar from adhering. Any portion of the dam that is left for a few days during construction in order to bring up other portions to the required height should not be allowed to dry, but water be constantly thrown upon its surface. The mortar should be as stiff as it can be conveniently spread, the joints all being well filled, and a specially selected spalls from the waste of the hardest stone from the quarry be firmly wedged in, so that the mortar is pressed and squeezed into all spaces round and about the joints and voids of the building stone. Grouting should never be used except with large cut blocks, when it is preferable to force the mortar into the joints with specially prepared implements.

In frosty weather all building must be stopped, as well as during heavy rains, as with frost the expansion of water in the mortar will disintegrate it, prolonged frosts necessitating sometimes the removal of the last layer built so as to secure a joint free from any suspicion of disintegration. Rain, on the other hand, immediately spoils the mixture of the mortar by washing away the cement covering the sand, and thereby destroying its adhesive power.

Stoppages caused by rains, frost, cessation of work on Sundays or holidays are likely to be productive of leakage along the joint of the fresh work. This can be obviated by the use of bags and tarpaulins as coverings, as a protection against rain and frost, and by not filling the joints between the stones with mortar—or, in other words, not bringing the work up flush each day. There is a minimum of area between the mortar which is set and the fresh mortar applied when worked in this way.

In placing large stones by the aid of a crane it is very necessary to leave ample width between them, as a slight touch from a heavy stone might break the already formed joint of the stone previously set in position.

It is extremely unsafe to use lime mortars for masonry dams, they being, unless eminently hydraulic, unable to resist any pressure of water. In some cases low dams have been built with lime mortar, with the face mortar containing a proportion of cement; the face joint being afterwards carefully pointed with cement and a specially selected fine sand. This practice cannot, however, be recommended for high dams, as the setting of the lime mortar requires a lengthy period, whilst it is very doubtful if the interior ever hardens. It may also happen that water is required to be accumulated in the newly formed reservoir immediately after the completion of the dam, which would be a highly risky proceeding if an unhydraulic lime mortar were used.

Some curious facts may be mentioned, not only to show the influence of a large body of masonry in retarding the solidification of the mortar in the interior, but also the danger of using rich limes in cases where such masses are necessary. Amongst them is the fact, cited by General Treussart, concerning the bastions erected by Vauban in the Citadel of Strasbourg in the year 1666. In the interior the lime, after 156 years, was found to be as soft as though it were the first day on which it had been made. Dr. John, also, mentions that, in demolishing a pillar nine feet in diameter in the Church of St. Peter at Berlin, which had been erected eighty years, the mortar was found to be perfectly soft in the interior. In both cases the lime used had been prepared from pure limestone.

As the site of the works for the formation of a reservoir is generally some distance away from any town or habitation, the cost of carriage may affect very seriously the cost of the scheme of water supply. Not unnaturally the district in the immediate neighbourhood of the site of the reservoir will be diligently searched for limestone of a quality to supply the cement required for the construction of the dam. A few rough tests may therefore be applied, should a limestone be found, to see if it be likely to furnish a hydraulic lime or cement. Such a stone, if serviceable, will generally have an earthy texture, and will weather to a brown surface. Acid, when applied in a few drops, will not cause such an effervescence as upon purer limestone. When breathed upon or moistened a clayey odour is emitted from the stone.

The best plan, however, is to burn a little in a small experimental kiln, and to afterwards observe the slaking and the behaviour of pats made from the paste.

The proportion of clay or other constituents in a limestone influences very greatly the setting properties of hydraulic lime without drying or the access of air.

Vicat, therefore, subdivided limes into the following three classes :—

Name of Class.	Percentage of clay associated with Carbonate of Lime only or with Carbonate of Lime and Carbonate of Magnesia.	Behaviour in slaking after being wetted.	Behaviour in setting under water.
Feebly Hydraulic.	5 to 12 per cent.	Pauses a few minutes, then slakes with decrepitation, development of heat, cracking, and ebullition of vapour.	Firm in 15 to 20 days. In 12 months as hard as soap — dissolves with great difficulty, and in frequently renewed water.
Ordinarily Hydraulic.	15 to 20 per cent.	Shows no sign of slaking for an hour or perhaps several hours—finally cracks all over, with slight fumes, development of heat, but no decrepitation.	Resists the pressure of the fingers in 6 or 8 days, and in 12 months as hard as soft stone.
Eminently Hydraulic.	20 to 30 per cent.	Very difficult to slake—commences after long and uncertain periods—very slight development of heat, sensible only to touch—very often no cracking or powder produced.	Firm in 20 hours—hard in 2 to 4 days—very hard in a month—in 6 months can be worked like a hard limestone, and has a similar fracture.

The method of testing of limes may be expedited by first taking a small basket full of the lime it may

be desired to test and immersing it in pure water from
six to ten seconds, allowing the uncombined water to run
off. Before cracking and falling to pieces begins, fill
a vase with the lime, pouring in water by the side of
the vase, that it may flow freely to the bottom, whence
it will be absorbed by the lime that is in an advanced
state of chemical action. Frequently stir and add
water, not to flood, but to bring the lime to the consistency of a paste; leave until the inert particles have
completed their action; this is announced by the
cooling of the mass, which may require three or four
hours or more. When all action has ceased beat up
again, and add water, if necessary, to produce the
consistency of potter's clay. Take a vase of this paste,
filling it even with the top, and immerse the vase in
water, taking note of the hour, day, month, and year
in which it was immersed.

From the above table it will be seen that an
artificial hydraulic lime may be made by moderately
calcining an intimate mixture of a fat lime with
as much clay as will give the mixture a composition like that of a good natural hydraulic limestone, of
which the product should be a successful imitation.

A soft material like chalk may be ground and mixed
with the clay in a raw state. Compact limestone, on
the other hand, requires burning and slaking in the
first instance—this being the most economical way of
reducing it to powder—then mixing with the clay and
burning a second time.

We have seen from the foregoing that lime mortar
should not be used, and that hydraulic limes may be

difficult to obtain in the district in which the works may be situated. It may consequently be assumed that imported cement will be used, and if that be the case all liability to doubt as to the permeability of the wall will be considerably diminished. It is therefore necessary to enter somewhat fully into the testing, manipulation, and porosity of various mixtures of cement with sand.

Portland cement differs very considerably in its characteristics and action. It can be manufactured more cheaply when under-burnt, because then a greater bulk of cement is produced with a given quantity of material, and it requires less fuel and less grinding; it also sets more quickly, but never arrives at the same ultimate strength as a burnt cement. Under-burnt cement contains, moreover, an excess of free quick-lime, which is apt to slake in the work and cause great mischief. This may be remedied by exposing the cement and allowing these particles to become air-slaked.

A slight difference in the manufacture may make a great difference in the character of the material, and rigid testing is necessary in order to secure the best cement.

Fineness of grit may be roughly tested by rubbing it between the fingers, or accurately by passing it through a sieve with meshes of known size.

The experiments of Messrs. Grant, Colson, and others show that when used neat a coarse grained cement is stronger than one finely ground; when mixed with sand, however, the finely ground cement makes stronger mortar than the other, the difference in its favour being greater as the proportion of sand in the mortar is greater. Where fineness of grit is alluded to in specifications, as it always should be, 14,400 meshes to the square inch is frequently specified, though the Metropolitan Board of Works used to specify, as well as a few engineers, that not more than ten per cent. by weight should be rejected by a sieve of 5,800 meshes to the square inch ; and there seems no doubt that this requirement, which is estimated to add only one-tenth to the cost of the cement, is a very desirable one to enforce. Good makers, however, generally grind their cement fine, and there need be no apprehension on this point. When the cement is obtained from recognised makers it is best to specify a cement that will pass a sieve with 14,400 meshes to the square inch, with not more than 10 per cent. residue, for reasons given further on.

Great care must be taken that finely ground cement is not lightly burnt, to prevent which the weight, or, better still, the specific gravity, of the cement should be tested. The weight of a cement is generally specified per striked bushel, it being considered that good weight per bushel is a sign of thorough burning ; but it is obvious that the weight is greatly influenced by the degree of fineness to which the cement is ground, also by the degree to which it has been aerated, and by the way in which the measure has been filled. The weight per bushel is therefore of little value.

The effect of fine grinding upon weight is shown

in the following results, obtained by Messrs. Currie and Co., of Leith :—

Meshes per square inch of sieve.	Percentage retained by sieve.	Weight of cement per bushel in lbs.	Weight of cement per cubic foot in lbs.
2,500	10	115	90
3,600	10	112	87
5,500	10	109	85
14,400	10	104	81
32,000	10	98	76

As the weight is therefore no reliable gauge of the quality of a cement, it is better to require a certain specific gravity to be given, which cannot vary with the different degrees of fineness of grit.

Mr. Grant, the Resident Engineer of the Metropolitan Main Drainage Works, found that the specific gravity of cement supplied by the best English manufacturers slightly exceeded 3·0, an inferior cement not being more than 2·8, whilst experiments showed the specific gravity of differently burnt cements to be as follows:—

Light burnt, 3·130.

Hard   ,,   3·134.

Medium ,,   3·131.

And his specification for specific gravity was not less than 3·1.

The specific gravity of cements can be easily ascertained by using Keate's Specific Gravity Bottle (Fig. F, next page), which is described by Mr. Grant in the Min. Inst. C.E., vol. LXII.

The bottle consists of two bulbs, the lower somewhat exceeding the upper in capacity. The exact capacity of the lower bulb is of no importance. On the neck between the bulbs is a file mark, *b*; on the neck of the upper bulb is a similar mark, *a*.

The capacity of the upper bulb between the marks
a and b must be accurately determined, and
may conveniently be either 500 or 1,000
grains, in water measure, at 60° Fahr.

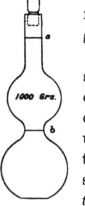

In ascertaining the specific gravity of a
solid in small fragments—small shot, for
example—the following is the mode of procedure : fill the bottle with distilled water
up to the mark b, accurately counterpoise
the bottle so filled in a balance ; drop the
substance *of which the specific gravity is to be
taken* carefully and gradually into the bottle

FIG. F.   until the water rises from b to a.   Ascertain
exactly the weight of the material so added.   If the capacity of the upper bulb be 1,000 grains of water the weight
of the material required to raise the water from b to a is its
specific gravity ; if the capacity of the upper bulb be 500
grains of water the weight of the substance added must
be multiplied by 2, which will give the specific gravity.

The principle of the apparatus is very simple.   The
capacity of the upper bulb is an exact measure of
distilled water, and when the water is raised from b to
a by dropping a solid into the bottle the bulk of that
solid, equivalent to the given volume of the distilled
water, is ascertained, and the relation between the
weights of the two is given by the weights of the
substances added, which is either the specific gravity
direct, if the capacity of the bulb be 1,000 grains, or it
can be ascertained by multiplying the weight of the
solid by the number which represents the part of 1,000
represented by the capacity of the bulb, &c.

The only precautions to be observed are that the air, which is apt to cling somewhat to the solid matter when dropped into the liquid, is carefully removed, and that if a very volatile liquid be used in the place of water the bottle should be stoppered or corked to prevent evaporation.

Besides the test for specific gravity, which, as has been pointed out, is of more importance than specifying the required pounds per striked bushel, other additional tests may be made which will give an indication as to some important qualities of the cement before using.

A thin glass bottle is filled with neat cement; if after some days it becomes set, and the bottle remains uncracked, it may be considered that the cement is not too hot. If the cement has shrunk within the bottle it is probably under-burnt; the shrinkage can be detected by pouring in a little coloured water. By filling a piece of glass tubing 18 inches long with neat cement, shrinkage can be very easily noted in a few days. Expansion and cracking may be tested by exposing a few well made pats of neat cement to hot steam, or by placing them in boiling water, after setting over night; five hours in hot steam of 200° Fahr. or over will be found sufficient to detect the slightest tendency to expand or crack.

All cements possessing a very high tensile resistance may be suspected of containing an excess of lime, and therefore, to prevent blowing, require considerable air-slaking or cooling.

In the gauging of cements great care must be

exercised. It is usual to specify that each cubic foot shall contain 87 to 90 lbs., and to obtain 90 lbs. per cubic foot some difficulties may be encountered in consequence of the extreme fineness to which cements are now ground; weight should, however, be insisted upon in the specification, to insure water-tightness. By using a conveniently shaped hopper with valve and spout, set at an angle of 45°, the required weight per cubic foot may be obtained; but it is necessary that the gauge box shall be of one cubic foot only—that is to say, measuring 12 inches square by 12 inches depth. If, for the purpose of saving time, double the quantity is used for twice the cube of sand, the gauge box must be increased by the square, the depth being constant. A box or vessel of 12 inches square and 24 inches deep will hold 14·2 per cent. more than twice the weight held by a cubic foot from settlement by its own weight, whilst gauge boxes of less than a cubic foot will contain proportionately less than the true equivalent of 87 to 90 lbs. per cubic foot.

It is advisable, from what has been previously mentioned, that the specified tensile strength shall not be placed too high, and it is better to require a moderate tensile strength, such as 300 to 350 lbs. per square inch, observing carefully the increased resistance to breaking by age.

Tests of cement mixed with sand to be used in the work are very desirable, but the length of time required, being not less than 28 days, renders it very difficult to carry out. It is also almost impossible to insure that the whole of the sand used be of uniform composition and quality as regards sharpness, surface of grains, &c. There can be no doubt, however, that this test will give more satisfactory information than can be obtained by testing the cement neat.

The test for tensile strength can be obtained by the use of a machine for that purpose, the moulds for making the briquettes being slightly rubbed with a greasy rag before putting in the paste. In making the neat paste not more than 20 to 25 per cent. in bulk of pure water must be used, and the pressure applied by the machine at the rate of 400 lbs. per minute. There are several classes of machines in the market, and in making a choice preference should be given to one that is conveniently geared to this speed. Generally there are difficulties experienced in working the machine at so slow a speed, the impulse being jerky when applied slowly by the hand. A difference of 25 per cent. can be obtained in tensile strength of neat cement between slow and quick speed, 800 lbs. being applied in less than one minute increases the true breaking weight by at least 25 per cent.

Tests of compression are also of great importance, but the apparatus required is cumbersome, and these tests are therefore rarely carried out. It may be generally assumed that resistance to compression is about twenty times that of the tensile strength.

The following table will indicate the tensile strength per square inch of Portland cement mortar of various mixtures as compared with the same class of cement neat :—

Age and time immersed.	Proportion of clean pit sand to 1 cement.					
	Neat cement.	1 to 1.	2 to 1.	3 to 1.	4 to 1.	5 to 1.
1 week.	445·0	152·0	64·5	44·5	22·0	
1 month.	679·9	326·5	166·5	91·5	71·5	49·0
3 months.	877·9	549·6	451·9	305·3	153·0	123·5
6 ,,	678·7	639·2	497·9	304·0	275·6	218·8
9 ,,	995·9	718·7	594·4	383·6	—	—
12 ,,	1075·7	795·9	607·5	424·4	317·6	215·6

The superiority of fineness to strength is fully shown from the following data, given by Mr. Grant :—

Age of Briquette.	Neat.		Three of sand.		Five of sand.	
	10·2 per cent. residue on a sieve of 2,580 meshes per square inch.	Sifted so as to pass all through sieve of 32,257 meshes per square inch.	10·2 per cent. left on sieve of 2,580 meshes per square inch.	All passed through sieve of 32,257 meshes per square inch.	10·2 per cent. left on sieve of 2,580 meshes per square inch.	All passed through sieve of 32,257 meshes per square inch.
Weeks.	Lbs. per square inch.	Lbs. per square inch.	Lbs. per square inch.	Lbs. per square inch.	Lbs. per square inch.	Lbs. per square inch.
1	353	346	75	252	31	136
4	533	380	171	330	97	208
8	585	469	206	358	118	223
25	710	495	282	397	166	272

The specification for the works of the Forth Bridge required that the cement should pass a sieve containing 50 divisions to the inch, equal to 2,500 meshes per square inch, leaving a residue of not more than 5 per cent. by weight. For the tests it was mixed with three times its weight of sand, which had been passed through a sieve of 400 meshes (20 divisions per lineal inch) and retained upon one of 900 meshes to the square inch (30 divisions per lineal inch).

About 10 per cent. of water was used in making the mortar. The briquettes were immersed in water after twenty-four hours, and so remained twenty-five days, when they were required to bear a strain of not less than 170 lbs. per square inch without breaking. For briquettes of neat cement the breaking stress after four days was not to be less than 200 lbs., and after seven days not less than 400 lbs. per square inch.

The class of sand selected will influence the strength of a structure very considerably, and may require more or less cement according to its quality. Pit sand has an angular grain and a porous, rough surface; it is therefore good for mortar. River sand is not so sharp, the grains having been polished by attrition, whilst sea sand is deficient from the same cause. Where great fineness is required, it should be ground and passed through a sieve. Clean sand should leave no stain when rubbed between the hands, but a sand so clean is rarely met with in up-country districts; washing will therefore be required in almost all cases, as the presence of clay and loam unfits it for all purposes. This can be readily done by stirring in a wooden trough having a current of water flowing through it.

Calcareous sands give stronger mortars than siliceous ones; sea sand, by containing salt, is apt to attract moisture, which can be largely obviated by washing in fresh water.

The water required to slake hydraulic limes or for the mixing of cement mortars varies very much, and is influenced by the district in which the works may be situated; in hot climates a large increase of water will be used. Hydraulic limes should be left after being wetted and covered up for a period of from twelve to forty-eight hours; the greater the hydraulic properties they possess the longer they will be in slaking; too much water will absorb the heat and check the slaking process. Strong hydraulic limes should be ground before using. The quantity of water required for mixing mortars will vary between 7 to 20 per cent. of the bulk of the ingredients; too much water must on no account be used.

In mixing the great object to be attained is to thoroughly incorporate the ingredients, so that no two grains of sand shall lay together without an intervening film of cement. On all works of over an estimated cost of £5,000 mortar mills should be adopted; this is absolutely necessary for the intimate incorporation of large quantities. The cement and sand should be mixed dry, the ingredients being turned over two or three times before the water is added; a very thorough incorporation of the materials is effected in this manner.

The bulk of mortar produced from various mixtures varies according to the size of the sand used. For estimating the quantities required, however, in a structure, three-quarters of the bulk of the cement *plus* the bulk of the sand in mixtures of 1 to 2 of sand to 1 to 4 of sand will be a near approximation to the total bulk obtained for building (see also page 77).

The artificial compound known as concrete is made by mixing lime or cement and sand with water and adding some hard material, such as broken stone,

clay, gravel, &c. The broken material is generally
called the aggregate, and the mortar which incases
it the matrix. Aggregates composed of angular fragments rather than rounded pieces (such as are obtained
from gravel) are to be preferred, their size not being
greater than will pass a ring of 2½ inches diameter
and not less than 1½ inch. The sand and cement
will be dry mixed by being turned over two or three
times before applying the water. When the mortar is
made it can be added to the aggregate, which has
also been carefully gauged and wetted, so that the
stones will not suck the moisture out of the matrix.
The whole, on being turned over three or four times,
will form a thoroughly incorporated concrete.

A very strong and impermeable concrete can be
made by a mixture of 1 of cement to 2 of sand and
3 of broken stone, the sand and cement together being
sufficient to more than fill the voids or interstices
of the aggregate. By the former rule one cubic yard
or 27 cubic feet of cement plus 54 cubic feet of sand
would, when mixed, equal 60·75 cubic feet of mortar.
Now experiment has demonstrated that a mixture
of 1 part of cement to 2 of sand and 3 of sharp
broken stone will equal about 0·60 of the total bulk
when thoroughly incorporated. We have therefore
$27 + 54 + 81 = 162 \times 0·60 = 97·2$ cubic feet, which is
greater than the cube of the aggregate by 16·2 cubic
feet. There is that amount, therefore, more than is
required to fill the voids in the aggregate. Care and
attention should be paid to these details, or a very
faulty concrete may be the result, the blame of which

cannot rest with the contractor, but must be borne
by the engineer. (See remarks upon porosity.)

The amount of water required for mixing will vary
between 10 and 25 per cent. of the total bulk of the
ingredients, according as the temperature of the air is
high, or low, or moist.

One precaution that must be taken is that an
absolutely clean water supply be obtained for the
mixing of the mortar and concrete. To thoroughly
wash the sand and stone used, so as to remove
clayey particles, and afterwards to use dirty water,
is equivalent to not taking any precautions whatever;
only a small proportion of clay in the sand or mortar
will literally kill out the adhesive qualities of the cement.

The concrete when made should be wheeled rapidly
to the place where it is required and gently tipped
into position, being rammed in layers of twelve inches
thickness; the layers should be horizontal, so that
there may be no trickling of water, which would carry
cement with it. Where surfaces are left by any
interruption or for convenience of work such surfaces,
before laying more concrete on them, should be swept
clean, made rough by a hand pick, washed, and
covered by a thin coating of cement. Generally a thin
milky exudation will be observed upon the surface
of the concrete last laid if a few days have passed.
This must be removed, as it will prevent the next
layer from adhering. A slow-setting cement is preferable to that which sets quickly, cracks being very
liable to appear after too rapid a setting. When set
the surface must be kept drenched with water, so that

the atmosphere may have no deleterious influence before further building operations are begun upon it.

We have seen that the fineness of cement increases the tensile strength of mortar; it has also a direct bearing upon the water-tightness of both mortars and concretes. This is a very important constructional matter in masonry dams, and merits attention. We have already noticed that mixtures of 1 of slaked lime or cement to 2 of sand, or 1 to 3 or 1 to 4 in volume, give about 75 per cent. of their total volume when mixed. This is, however, but a rough average and indication of the water-tightness of the cement mortar. As it is evident that 1 to 2 will give, by this calculation, 2·25, and therefore contains an excess of cement, 1 to 3 will give 3, the voids being just filled, whilst 1 to 4 gives 3·75, or 0·25 less than the original sand, showing that there are still voids to be filled. By mixing 1·32 in volume of cement to 4 of sand we have the original 4 of sand with the interstices filled.

Investigations made by the author in a series of experiments have led to the following facts being disclosed. An up-country sand obtained from rivers, when sifted through a sieve with one-eighth inch square holes, and afterwards washed, will contract when thoroughly saturated with water, as it will be in mixing, 23·3 per cent., and the voids remaining are equal to 32 per cent.—that is to say, that wet sand, as taken from the washing boxes or troughs and placed in the gauge box, will afterwards shrink, when mixed by the action of water and movement,

23·3 per cent., there still remaining 32 per cent. of voids to be filled with the cement. This affects, it will be seen, very considerably the calculated quantity of sand required in works of large dimensions, as the total washed sand will only equal, when used in the building, 76·7 per cent. of the gauged quantity.

We should have, therefore, in a mixture of 1 of cement to 2 of sand by gauge, 68 per cent. of the total volume as mortar; in 1 to 3, 64 per cent.; and in 1 to 4, 62 per cent. of the total as mortar. Now water-tightness is obtained by filling the voids in the sand, and as the sand, when thoroughly saturated with water, as occurs in making the mortar, contains 32 per cent. of voids, it is easy to see that 1 of cement to 3 of sand is about the limit of sand that can be given if water-tightness be required, there still being a slight excess of cement over and above the voids contained in the sand. As the sand, however, when gauged, will shrink in mixing 23·3 per cent., we have for the original 3 only 2·30 in volume of sand, which, containing 32 per cent. of interstices, results in 1·56 of solid sand, which, when mixed with 1 of cement, equals 2·56 of mortar. It will also be seen that when the 1 of cement is added to the 2·30 actual of sand the voids are more than filled, as 2·56 results in volume, or 0·26 more of cement than is required to fill the interstices. This excess is required for the adhesive films between the building stones and the mortar. By calculation, then, we are assured that the construction, when built of this mixture, will be water-tight.

As a demonstration that the above fulfilled all requirements, a masonry dam was constructed in the Province of Huelva, Spain, under these conditions, the mortar being composed of 1 of cement to 3 of sand. A little more than one-third of the whole structure was mortar, the remaining two-thirds being stone ; the dam having a maximum height of 70 feet. The mortar was exposed to a water pressure at the base of 30 lbs. per square inch, which it resisted, the down stream face being perfectly water-tight after sweating slightly for the first few months.

A water-tight concrete can be made by mixing 1 of cement to 2 of sand with 3 of broken stone not larger than $2\frac{1}{2}$ inch cube or less than $1\frac{1}{2}$ inch cube. Here it is seen that the sand becomes 1·53 with 32 per cent. of voids, or 1·04 of sand to 1 of cement, the mixture being 2·04, or 0·51 excess of cement over and above the voids of the sand. The 3 of broken stone with the 2 of sand and 1 of cement will equal, when mixed, 60 per cent. of the total volume, or 3·6. The voids in the broken stone are therefore 52 per cent., and, as 52 per cent. of 3 equals 1·56, the 2·04 of sand and cement are more than sufficient to fill the voids by 0·48. There is consequently ample cement in this mixture to insure water-tightness.

The above figures are obtained, as before stated, with sand that was riddled with a one-eighth inch square hole sieve, and will, of course, vary with the quality of the sand used. The cement was finely ground, 90 lbs. per cubic foot being obtained with difficulty, owing to its fineness. Eighty-five pounds per cubic foot in a mixture of 1 to 3 of sand would hardly be sufficient with the above class of sand to fully occupy the interstices; porosity would therefore not be overcome. For the purposes of estimating costs and demonstrating water tightness they may with safety be adopted.

It will be observed that the volume of cement in the above mixtures is accepted as such without any allowance for shrinkage. Ninety pounds per cubic foot of finely ground cement contains about 50 per cent. of voids; the actual cube is consequently only half, but the addition of the water for mixing and the pores left in the mortar when mixed and set make up the difference to within a very small percentage. For simplification of calculation and deduction the cement has been retained as a full volume in all cases. A hand-made mortar will always contain a few air holes or voids when set unless very well mixed. A 1 to 3 mortar when set, if broken across and examined under a magnifying glass, will be found to contain more open spaces or voids than a 1 to 2 mortar when examined in the same way, the few pores observable being partially attributable to air in the mixture. Well made machine mortar and concrete is very free of air holes or voids when subjected to the same examination.

The mortar used in the construction of the Vyrnwy Dam, for the supply of water to Liverpool, was 1 part of cement to 2 parts of sand. Experiments were made as the work progressed, and it was found that by pulverising the stone—a grey slate rock, of which the dam was built—and mixing 2 parts of this with 1 of the sand, and using 2 parts of this mixture with one of Portland cement, a stronger mortar resulted, and this mixture was ultimately adopted.

The Geelong Dam, Victoria, Australia, was built entirely of concrete, being 60 feet deep. It was subjected to a maximum water pressure of 26 lbs. per square inch. The best results were obtained by mixing the ingredients in the following proportions :—

2-inch stone	....	....	$4\frac{1}{2}$	parts.
Screenings	....	....	$1\frac{1}{2}$	,,
Sand	....	....	$1\frac{1}{2}$	,,
Cement	....	....	1	,,
Total	....		$8\frac{1}{2}$	parts.

In this case the 2-inch stone and screenings when mixed could not have contained more than 30 per cent. of voids, as the $1\frac{1}{2}$ of sand and the 1 of cement would not give much more than 1·78 of mortar, and in this way about 70 per cent. of the original cube would be obtained for building. There is certainly no excess of cement in this mixture. The average weight of the concrete was 143 lbs. per cubic foot.

When the reservoir was filled a little water found its way through the dam, but this leakage soon stopped, owing to hard incrustations of lime being formed.

The Tytam Dam, near Hong Kong, was built with three classes of concrete and ashlar facing of granite ; behind the ashlar masonry there is two feet of fine concrete, composed of 4 parts of stone (1-inch cubes), 6 parts of sand, and $2\frac{1}{2}$ parts of Portland cement, which is 5·63 parts of mortar to 2·08 of voids in the stone, giving a great excess of mortar. This formed a water-tight skin. Next came five feet of concrete, composed of 4½ parts of stone, 3½ parts of sand, and 1 of cement, the hearting consisting of the last class of concrete with stones of 3 to 6 feet in size embedded in it. These stones were kept well apart, so as to allow the concrete to be well rammed between them. The ashlar masonry was grouted with mortar composed of 1 of cement to 2 of sand.

The height of this dam is 95 feet; the maximum pressure of water is therefore 41 lbs. per square inch. Any water that might leak into the wall was allowed to escape through perforated zinc pipes 1½ inch diameter, which were placed 5 feet apart. When the water had risen 45 feet the leakage could be carried off by a 1-inch pipe without pressure; it was therefore very slight with that depth of water.

Mr. J. Watt Sandeman, in an excellent and useful paper contributed to the Min. Inst. C.E., vol. cxxi., on Portland cement and concretes, gives the ratio of interstices of different aggregates as follows:—

	Weight of aggregates per cubic foot.	Ratio of interstices.
	Lbs.	Per cent.
Broken limestone, the greater part of which would be gauged by a 3-inch ring	95	50·9
Gravel, screened free from sand, varying in size between small pebbles and pieces gauged by a 2½-inch ring	111½	33·6
The above limestone and gravel well mixed in equal proportions	113½	33·6
Sandstone, varying in size between pieces gauged by a 4-inch ring and pieces gauged by an 8-inch ring	74	50·0
Sandstone, varying in size between sand and pieces gauged by a 4-inch ring	92	34·0
The two above sandstones mixed in equal proportions	91¼	36·0

And he further remarks, on the manipulation of concretes, that, from studying the facts adduced in regard to the greater strength and economy attained by duly proportioning the components of concrete, he strongly deprecates the slovenly and unscientific practice of using in the manufacture of concrete unscreened gravel and sand, as although in cases where such material is employed the approximate ratio of sand to gravel is sometimes supposed to be ascertained by sifting a few samples, yet this ratio varies so widely— viz., between 19·5 per cent. and 113·6 per cent., according to the tests of Mr. Colson—that it would be quite impossible to obtain uniformity. Moreover, as the ratio of sand to gravel in the most economical concretes has been shown to range between 28½ per cent. and 45 per cent., it is clear that the varying natural ratio of sand to gravel would not make economical concrete, as, even assuming that the ratio of sand in any gravel were exactly according to requirements, it would be immediately altered by casting the material into waggons or barges. Again, by tipping them into a heap the sand and gravel always separate and assort themselves under the influence of gravity, and the author has witnessed the practical impossibility of combining the two again in due proportions when shovelling them out of a heap. Where concrete is made from unscreened gravel there must of necessity be portions of it in which the interstices of the gravel are not filled, and others in which the mortar is in excess, and in consequence the strength of such concrete is very variable. It would also be quite impossible with unscreened materials to make water-tight concrete, except by using an unnecessary excess of cement, entailing a proportionately increased cost.

In hand-mixed concrete the quantity of water necessary to render one mixing of the proper consistency having been noted, the same quantity of water should be used for each mixing, the uniform consistency of the concrete being of vital importance. The thorough hydration of the cement is also of paramount importance; but, as the quantity of water required will vary with the age and quality of cement, the character of the sand and the aggregates, and the state of the weather, the proper quantity to insure thorough hydration, without washing the cement out of the concrete, or causing the aggregates to settle down in contact with each other, can only be determined by the observation of an engineer experienced in the manufacture of concrete.

It may be stated as a guide that, when deposited *in situ*, the concrete should be somewhat of the consistency of dough, and should always yield sufficiently to allow of a man treading it to sink in it to a depth of at least six inches. Workmen should always be employed to tread and shovel each mixing of concrete throughout while it is being deposited; otherwise water-tightness will not be attained. In order to insure this engineers should require contractors to provide men treading the concrete with waterproof boots.

Concrete mixing should not be allowed to proceed during very windy or wet weather, unless performed under efficient shelter, as a considerable percentage of the finest and most valuable part of the cement will be blown away, and during rain the cement and sand cannot be efficiently mixed.

The concrete should in all cases be deposited in such a manner as not to allow the aggregate to separate from the mortar under the influence of gravity. To lessen this risk the size of the largest aggregate should be limited to such as would pass through a sieve with meshes two inches square. Further, to prevent the separation of aggregates, concrete should never be passed down a shoot; and, if tipped from a height, a layer of two to three feet in depth should first be deposited by lowering it in tubs. For large works regularity in mixing concrete may be attained by the use of a mechanical mixer, provided it be an efficient one, as there are so-called mixers worse than useless by causing the separation of the aggregates from the mortars. The chief advantage of a good machine is the facility which it affords for making concrete with a uniform quantity of water and of uniform consistency, which cannot be easily attained by hand-mixing.

Having noted the essential points as regards the stone, mortar, and concrete, we now come to another important part of the construction—that is, the proportion of mortar to stone in the whole structure. A heap of tipped building stone, without waste, will have about 50 to 52 per cent. of voids, whilst the selected stone and its waste, or the whole of the material won by the quarrying operation, if tipped apart, will augment in volume about 33 per cent. These figures are mentioned as showing that not more than one-third of the whole structure should be mortar, as otherwise the masons are not obtaining less voids around the stones than would be given by tipping the stones roughly on the ground. By laying the stones carelessly, therefore, an excess of mortar would be required, which would not only expose the structure to additional sweating, but increase the cost very considerably. If any leakage result it will be through the mortar or joints rather than through the stone. The stones must not only be carefully laid, but a large quantity of spalls should be driven in to fill all corners, points, and voids in and around the building stones; and there is no doubt that with skilled labour 0·75 in volume of mortar to 2·25 of stone can be attained in rubble masonry. But generally these structures are built up-country, and far removed from the probability of obtaining skilled labour; 1 of mortar to 2 of stone is therefore nearer what will be obtained in practice under skilled supervision.

It is generally thought that in the thorough setting and ageing of the mortar, in masonry walls of such thicknesses as are required to resist the pressure of water in even moderately high dams, they become a perfectly solid and immovable block, but such is not the case. This points, therefore, to the necessity of most careful supervision during construction; otherwise the calculations applied to the form of the adopted profile will be erroneous, the least amount of masonry being used that will bring the lines of resistance within the middle third of the profile. This, although leaving a margin of safety, nevertheless requires the dam to be well built. The Vyrnwy Dam has a bore-hole connecting the sill to a gallery running longitudinally through it rather more than 80 feet in depth; in the bore-hole is suspended a steel pianoforte wire carrying at its lower end a heavy weight immersed in water; near the bottom of the wire is a seismograph, the movements of which are multiplied four times. The first observations were made when the water in the reservoir was within 13 feet of the sill, or top of the dam; during this 13 feet rise of water the sill moved horizontally, in relation to a point 80 feet below, to the extent of 0·868th of a millimetre. The changes of temperature caused the dam to move sensibly from day to night and from night to day, whilst earth tremors were clearly recorded. The dam faces south-east, so that it does not receive the full heat of the sun; but when the reservoir is full the difference between a hot summer day and night was 0·366th of a millimetre, due to the expansion of the outer face of the wall.

We have here the first and only recorded movements of a dam from the above various causes—which took place, it must be remembered, in a wall that was increased considerably in width over and above the required calculated profile from about one-third of its height from the top to its base, as the whole crest or top of the wall was utilized as a weir over which the surplus water from storms, &c., might pass.

It is therefore evident that too much care and supervision cannot be bestowed upon the workmanship and materials in the construction of high dams.

However carefully a dam may be built there will always be a certain amount of leakage, which should not amount to more than sweating on the down stream face when the reservoir is first filled. This loss of water in well built dams disappears after a few months. What is more serious, and will give cause for grave anxiety and trouble, is the appearance of leakage from the base and foundations of the wall. This, if not great, will also seal in course of time, but it is certainly advisable to make careful observations as to the volume given and its variation with the level of the water in the reservoir during a period of several months. If the leakage be permanent, the volume will vary in proportion to the square root of the head of water in the reservoir.

Such a leakage involves sometimes an immense amount of costly work, and generally necessitates the emptying of the reservoir before anything effectual can be done. If the leakage be suspected to be at the lowest point, or near the lowest point, of the dam, the basin of the reservoir that is in close proximity to the wall may be covered with a well prepared clay, which can be tipped into the water and allowed to settle by its own gravity round and about the inner base of the wall, the pressure of the water tending to drive the clay into

the false joints or vein. This operation is necessarily
done with the level of the water lowered to a convenient
depth. The water for a short period will be made turbid
from a small portion of the clay being in suspension,
but this will soon fall, and the water will again become
clear. A little alum thrown in will clear the water
very quickly, and may with advantage be employed in
case of urgency.

The above procedure is applicable in cases of dams
of considerable height in closed valleys or gorges, where
the inner toe can be easily protected and covered. An
extensive and shallow basin would necessitate other
methods being adopted, such as a trench being cut in
the solid rock along the inner toe of the wall, to be
afterwards filled with puddled clay or cement concrete,
in either case being connected to the wall on the inner
face in such a way that the wall and the material of
the trench form a complete union. This will form an
apron, and most effectually cut off any possible leakage,
but is an expensive and tedious operation.

Another danger, already mentioned, to which a wall
is exposed, particularly along its top, is that of expansion and contraction, and although this is allowed for
by the formula for the profile, still the greater the
length of the top the greater the danger becomes of
vertical cracks appearing in the course of time, particularly if during the heat of a dry season the reservoir
be only partially filled. The top will, in this case, be
exposed to considerable variation, and possibly result
in a few cracks of slight thickness appearing, but
nevertheless of sufficient size as to permit a consider-

able quantity of water to pour through the wall. Between the heat of summer and the cold of the winter months the top of a wall will be exposed to a variation in length equal to about one inch on every 350 feet length when built of rubble masonry or concrete. The interior of the masonry of a wall is not exposed to the same variation as the exterior, but nevertheless the influence is greater than might be expected, as demonstrated by an experiment made in America, where the range of temperature in a wall of five feet six inches was, during twelve months, as registered by a maximum and minimum thermometer, about 20°. The face of a dam may, from the same cause, be cracked in many directions, an instance of which is the concrete reservoir wall at Colombo, which is fissured in all directions, and is now shielded from the sun's rays by an earthwork slope.

Vertical fissures appearing through a wall do not necessarily endanger the structure when caused by contraction, and they may be readily calked by oakum derived from old ropes. This will be slightly elastic; and for some few years yield with the movement of the wall. Grouting should not be employed on any consideration, neither is pointing on the inner face effective or desirable.

As appertaining to the question of leakage of masonry dams, the following interesting pages are appended, relating to the investigation of the causes of failure of the Bouzey Dam, in France.

THE Bouzey Dam is situated about 4½ miles from the town of Épinal (Vosges) ; its general dimensions are—

Length on top	....	1,700 feet.
Height above river bed	....	49 ,,
,, ,, foundations	....	75½ ,,
Width at top	....	13 ,,
,, ,, base	....	39 ,,

The plan of the wall is straight. The profile was designed to be one of "equal resistance," each joint being assumed perpendicular to the resultant of all the forces acting on it. When the reservoir is full there is tension in the masonry near the back face.

The dam was founded on a porous conglomerate rock consisting of silicious stones joined together with a rather weak cementing material. Specimens of this rock resisted a crushing load of 274–548 tons per square foot, but broke under a tensile strain of only 10·10 tons per square foot.

In the Minutes of the Proceedings of the Institute of Civil Engineers, vol. cxxv., 1895–96, Pt. 3, page 461, an abstract is given from the report of a special commission of the Ponts et Chaussées upon the failure of this dam, from which the following is taken, as the investigation made and conclusions formed are of great practical importance to all designers of masonry dams :—

"Mr. Delocre assumes in his calculations, taking a unit length of the dam, that, in any horizontal section, the thrust due to the horizontal pressure of the water above that level is counteracted by the friction and cohesion of the mortar in that plane, and that the section has to support a vertical pressure equal to the weight of the portion of the dam above that level applied at the point where the resultant of that weight and the water pressure intersects the plane. The intensities of the pressures are calculated on the assumption that the stress varies uniformly across the section. The profiles of the Furens Dam (164 feet high) were calculated according to this method, the maximum vertical pressure being limited to 5·9 tons per square foot; likewise those of the Ban Dam (148 feet high), but in this case the limiting pressure was increased to 7·3 tons per square foot. In the Tournay Dam (116 feet high) the limit was taken as 6·4 tons per square foot.

"After the completion of the last-mentioned dam it was decided to raise it 3·3 feet, and Mr. Bouvier, in recalculating the stresses, introduced a modification into the method adopted by Mr. Delocre, by which the maximum pressures arrived at by the first system are divided by the cosine squared of the angle which the resultant makes with the vertical to get the true maximum. This gave the maximum pressures in the Tournay Dam as 8·5 tons per square foot instead of 6·4 tons per square foot by the first method ; and he introduced that, taking account of this correction, a dam built of good hydraulic lime might be subjected to a maximum pressure of 9·0 tons per square foot after two or three years and 12·8 tons per square foot after twelve years.

"Mr. Guillemain, while adhering to Mr. Bouvier's method, advocates that the pressure should be calculated on all planes passing through the point sustaining the maximum pressure, so as to arrive at the actual maximum pressure in a dam.

" The effect of a dam being waterlogged below a certain level for a given distance from the face is next considered, and is shown to decrease the amount of the resultant and to move its point of application nearer to the outside face.

"The figure (Fig. G, p. 94) shows the section adopted for the dam in dotted lines ; it was approved of by the Council of the Ponts et Chaussées on the understanding that the dam should not be raised at once to its full height, but that the level of the water should be kept provisionally at 1,212·0 feet above datum, and not be raised to its ultimate level, 1,218·5 feet, until the dam had thoroughly consolidated. The lowest draw off was fixed at 1,169·3 feet. The dam was straight, and its length was 1,700 feet ; the contents of the reservoir at the provisional level being 1,034 million gallons, and at the final level 1,540 million gallons. The dam was built on the new red sandstone, which was fissured and permeable, and to prevent the water getting underneath the dam a guard wall was carried down beneath the water face into the

solid rock. The masonry was built with hydraulic
lime mortar, and the inner face coated with plaster of
cement mortar 1½ inch thick.

"The guard wall was built in 1878 and 1879, and
in its construction springs were met with that were
with difficulty sealed. The dam itself was built up

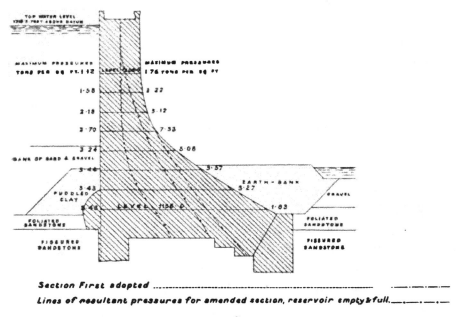

Section First adopted .... —·—·—·—·—

Lines of resultant pressures for amended section, reservoir empty & full. —·—·—·—·—

FIG. G.

to the level of 1,187 feet for a length of 820 feet in
1879, and completed in 1880. During its construction the engineers, on representing to the Minister
of Public Works the greatly increased storage that
would be obtained at a small additional cost by at
once raising its height to the ultimate level, obtained
permission to do so.

"The filling of the reservoir was started in November,

1881, about a year after the completion of the masonry, with the water of the river Avières. When the water in the reservoir attained the level 1,187 feet springs appeared on the lower side of the dam having a flow of about two cubic feet per second. In December, 1882, two fissures, at distances of 243 yards and 332 yards respectively from the overflow, were noticed in the dam, and supposed to be due to changes in temperature, as they were also apparent during the preceding winter. These caused the discharge of the springs to increase to 2·6 cubic feet per second. By the 4th December, 1883, the reservoir had only been filled up to a level of 1,197 feet, but after that date the filling proceeded at a greater rate, owing to the utilization of the waters of the river Moselle.

"On the 14th March, 1884, when the level of the water in the reservoir was 1,210 feet, a length of 444 feet of the dam suddenly assumed a bent form between the points 119 yards and 267 yards, and the flow of the springs increased from 2·6 cubic feet per second to 8·1 cubic feet per second. The height of the water in the reservoir was kept constant for a year after this. No further movement took place, and the flow of the springs remained nearly the same.

"In 1885 a bore-hole was put down on the lower side of the dam, and afterwards the reservoir was emptied to ascertain what had actually happened. It was found that the dam had separated from the masonry wall beneath it between the points 148 yards and 247 yards, still keeping the vertical, the greatest deviation from the straight being 1·1 foot at the centre of deflection. On the inner face, at each end of the displaced length, was a group of fissures. That on the right on the inside, only visible on that face, consisted of three cracks at the points 117 yards, 123 yards, and 130 yards from the overflow respectively, descending obliquely to the base of the dam, the last one joining the horizontal cracks by a ramification. On the left were four cracks visible on the inside face—the temperature crack at 243 yards, which went vertically through the dam to the outside face, as also did the crack at 267 yards. The crack at 256 yards was vertical, but visible only on the inner face, and finally a fissure, inclined at about 45°, at 267 yards, which cut at its base the fissure at 256 and joined the horizontal fracture. The cracks at the centre of the deflection were visible on the outer face.

" The formation beneath the displaced part of the dam was dislocated for two or three yards in depth, and two crevices were noticed from which springs issued, also deposits of clay, in lenticular beds, which usually were less than one-tenth inch in thickness; above the dam fissured and permeable beds were found which passed beneath the foundations of the guard wall.

" Following the recommendations of a special commission of the Ponts et Chaussées, it was resolved to form an abutment of masonry on to the solid rock at the outer toe of the dam, starting at the level 1,182·4 feet, drains being laid through the masonry to lead away any water percolating underneath the dam. A wall of masonry was to be built on the water side at the junction of the guard wall and the masonry of the dam proper, and covered with puddled clay to a depth of about three yards. Wherever the cement plastering had become detached the joints were to be raked out and filled with cement mortar. The fissures also were to be filled with cement mortar, or cement grout when difficult of access.

" The tubes lining the boring on the lower side of the dam were carried up above ground level, so that as the water rose in the reservoir its height in this tube could be gauged.

" These works of repair were executed in 1888 and 1889, being complete on September the 14th, 1889; the figure gives the amended section in full lines. The filling of the reservoir was recommenced on the . 18th of November, 1889.

" The flow of the springs increased from 0·5 cubic feet per second, to start with, to 2·8 cubic feet per second on the 15th of May, 1890, when the level of the water in the reservoir was 1,218·5 feet above datum, that in the tube lining the bore-hole being 38·4 feet lower. The level of the water in the tube followed that in the reservoir as the latter filled.

" The deflection of the two points distance 193 yards and 267 yards from the overflow, the middle and one end of the portion displaced in 1884, was read by means of a theodolite.

" The reservoir began to be used for feeding the Canal de l'Est on 15th May, 1890. The level of the water was raised each year to 1,218·5 feet, and was never lower than 1,210·6 feet. The flow of the springs varied from 1·4 cubic foot per second to 2·6 cubic feet per second. The maximum level of the water in the tube was 1,184·4 feet, and it was never lower than 1,179·5 feet.

" The point at 193 yards deflected 0·32 inch to 0·72 inch, and the point at 267 yards from 0·04 inch to 0·25 inch. The vertical fissures opened in the winter and closed in the summer, attaining a maximum width of 0·28 inch. On April 27, 1895, at a quarter to six o'clock a.m., when the level of the water in the reservoir was 1,218·2 feet, a length of 594 feet of the central part of the dam was suddenly overturned at the level 1,186 feet, between the points at the distances 149 yards and 347 yards respectively. This length includes all but 90 feet of the part that was displaced in 1884. The fracture was nearly level longitudinally, and transversely it was level for 12 feet and then dipped towards the outside. The foundations had not moved, and the masonry was found to be of good quality.

" The commission appointed to inquire into the cause of the disaster calculated, according to Mr. Bouvier's formulas, the maximum intensities of pressure in the dam for the provisional and ultimate height of the water in the reservoir, and for an additional elevation in its level of 1·6 foot to allow for the highest possible overflow level; and the results are given in the accompanying diagram for the ultimate height of the water. The weight of the masonry was taken at 125 lbs. per cubic foot. With the ultimate height

of the water at the level 1,182·4 feet, this method of
calculation gave a maximum intensity of pressure
of 7·3 tons per square foot. and a ratio of water
pressure to weight of masonry above that point of
0·695; with the extra elevation of 1·6 foot, a maximum pressure of 10·6 tons per square foot and the
above ratio of 0·736.

"Mr. Bouvier gives the maximum intensity of
pressure in five dams as follows :—

Name.	Date of construction.	Height of water in reservoir.	Maximum stress in masonry.
		Feet.	Tons per square feet.
Gouffre d'Enfer ... ...,	1861—1868	164	8·6
Tournay, before rising	1861—1867	116	8·5
„ after rising 3·3 feet ...	1861—1867	119	11·0
Ban ... ...,	1866—1870	148	10·0
Pas-de Riot...	1833—1878	110	9·1
Chartrain ... ...,	1888—1893	151	9·4

" Thus the maximum pressure in the amended dam
at the level of 1,182·4 feet, calculated in this manner,
was less than in the above five dams. But an examination of the diagram showed that the resultant fell
outside the middle third of the section. Therefore, if the
dam below this level (the point at which strengthening
began) was considered as fixed, and the stresses worked
out in the usual way for the joint at the level of 1,182·4
feet, the maximum vertical pressure thus found was
4·6 tons per square foot, and the maximum tension 1·3
tons per square foot ; and, for the joint 3·3 feet above,
the maximum pressure and tension were respectively
4·3 tons per square foot and 1·0 ton per square foot.

" The cracks in the dam had not been properly filled

with cement grout, as had been proposed, because they were for the most part very narrow and would not take it, but they had been closed on the face by means of tarred yarn fixed with wooden wedges. This left the interior of the fissures still open. The vertical fissures were not harmful, but the one at 267 yards communicated with the oblique crack, and a dangerous uplift was caused at that point. If the fissure had a depth of five feet the maximum pressure in that section would be 18·1 tons per square foot ; if it had a depth of ten feet the resultant pressure would be outside the dam at this section. It thus appeared that the oblique fissure at 267 yards could determine the rupture of the dam for some yards in length, unless the support this portion received from the adjoining parts kept it in position. The results stated above showed that the remainder of the dam was in a state of tension on the inside face at the joint at a level of 1,182·4 feet and for some distance above. In the first season's work the dam was built up to the level of 1,187·0 feet, and about 18 inches was taken down on restarting work the next season ; consequently a weak place at the junction of the two seasons' work occurred at the same section in which tension existed. It appeared that the dam gave way at the oblique fissure first, and brought the remainder of the 594 feet, which was in a state of excessive strain, with it. This coincided with the evidence of the only eye-witness and with the state of the dam after the accident.

" The commission agreed that if the provisional level for which the dam was approved had been adhered to

no serious accident would have happened, as in that case there would have been no tension. The following are the conclusions of the commission :—

"1. The masonry of the Bouzey Dam was exposed to tensions which exceeded its powers of resistance on account of the defective adhesion of the part built in 1880 to that built in 1879.

"2. The catastrophe of Bouzey shows that it is necessary to so design reservoir dams that the masonry is not exposed to any tension.

"3. In case of such accidents as that at Bouzey in 1884 there should be no hesitation in rebuilding entirely all portions of the masonry in which there are fissures which might give rise to uplift.

"4. The conditions of stability of existing dams should be inquired into, and, if necessary, the level of the water reduced so as to do away with all tension in the masonry."*

* The failure of this dam is often referred to by engineers, and a considerable amount of thought and labour has been given to demonstrating the reason of its failure, without indicating, however, the reason of its having stood so long, which was indeed astonishing considering its profile.—C. F. C.

EARTH tremors and quakes are likely to produce disastrous effects on masonry dams in which there is no elasticity. Earthen dams, on the other hand, particularly if not of great height, will resist very considerable movement of the earth without damage; the lashing backwards and forwards of the water of the reservoir being the greatest danger from which they are likely to suffer.

Professor John Milne, in an appendix upon this subject written for Mr. W. K. Burton's work on the "Water Supply of Towns," says :—" That great advantage can be gained by selecting a proper site in earthquake countries is indicated by the fact that in building regulations for districts where earthquakes are frequent—as for example in Manilla and Ischia—special references are made to the methods of founding on soft ground, and in many cases certain areas have been marked off as unsuitable for buildings of any description."

He further states "that in making dams for impounding reservoirs it must not be forgotten that these may be subjected to a horizontal shaking, with the result, if they are made of loose material, that this may be disintegrated much in the same way as a pile of sand would be on a shaking table. For this reason it is better in earthquake countries to follow the English rather than the American practice in designing dams for impounding reservoirs.

"In connection with these dams it must be remembered that they may possibly be topped by waves caused by the earthquake motion. This is a strong argument for the adoption of a dwarf wall as a 'breakwater' along the top of an earthwork dam in an earthquake country. It might be further advisable to pitch with stones the top and lower slope of the dam, as well as the upper slope, so that if it were merely a case of topping with a few waves in succession there would not be the danger of cutting away the earthwork that there would be without the pitching."

It is impossible to give any rule which might be followed with safety, but one can only point out that large impounding reservoirs found in earthquake countries are a source of constant danger to any town or village that may be situated below, and in the direction which the freed water would take if a rupture were caused by earth tremors. A series of small dams, on the other hand, would resist very effectively any moderate quake of such force as would completely destroy an embankment or wall of great height.

APPENDIX I.

18,000 cubic yards of masonry = $\left\{ \begin{array}{l} \text{12,000 cubic yards of stone.} \\ \text{6,000 cubic yards of mortar.} \end{array} \right.$

6,000 cubic yards of mortar = $\left\{ \begin{array}{l} \text{2,344 cubic yards of cement.} \\ \text{7,031 cubic yards of sand.} \end{array} \right.$

2,340 cubic yards of concrete = 3,900 cubic yards of aggregate and cement = $\left\{ \begin{array}{l} \text{650 cubic yards of cement.} \\ \text{1,300 cubic yards of sand.} \\ \text{1,950 cubic yards of stone.} \end{array} \right.$

The above concrete and mortar can be further subdivided to obtain the cost per cubic yard for each class.

Mortar :

	£	s.	d.
1 cubic yard of cement at 90 lbs. per cubic foot = 2,430 lbs. at 40/- per ton =	2	3	5
3 cubic yards of washed sand at 4/- = ...	0	12	0
4 cubic yards × 0·64 = 2·56 cubic yards =	£2	15	5
or 21/8 per cubic yard ; therefore			
1 cubic yard of mortar = ...	1	1	8
Mixing and carrying to site = ...	0	2	0
Water pipes, &c. = ...	0	0	4
Total per cubic yard	£1	4	0

APPENDIX I.

ESTIMATE OF COST FOR MASONRY DAM 100 FEET IN HEIGHT, SHOWING SUBDIVISION OF THE MATERIALS.

18,000 cubic yards of masonry $=$
$\begin{cases} \text{12,000 cubic yards of} \\ \quad \text{stone.} \\ \text{6,000 cubic yards of} \\ \quad \text{mortar.} \end{cases}$

6,000 cubic yards of mortar $=$
$\begin{cases} \text{2,344 cubic yards of} \\ \quad \text{cement.} \\ \text{7,031 cubic yards of} \\ \quad \text{sand.} \end{cases}$

2,340 cubic yards of concrete $=$ 3,900 cubic yards of aggregate and cement $=$
$\begin{cases} \text{650 cubic yards of} \\ \quad \text{cement.} \\ \text{1,300 cubic yards of} \\ \quad \text{sand.} \\ \text{1,950 cubic yards of} \\ \quad \text{stone.} \end{cases}$

The above concrete and mortar can be further subdivided to obtain the cost per cubic yard for each class.

Mortar :

	£	s.	d.
1 cubic yard of cement at 90 lbs. per cubic foot $=$ 2,430 lbs. at 40/- per ton $=$	2	3	5
3 cubic yards of washed sand at 4/- $=$ ...	0	12	0
4 cubic yards $\times$ 0·64 $=$ 2·56 cubic yards $=$	£2	15	5

or 21/8 per cubic yard ; therefore

	£	s.	d.
1 cubic yard of mortar $=$	1	1	8
Mixing and carrying to site $=$	0	2	0
Water pipes, &c. $=$	0	0	4
Total per cubic yard	£1	4	0

Concrete:

	£	s.	d.
1 cubic yard of cement at 90 lbs. per cubic foot = 2,430 lbs. at 40/- per ton =	2	3	5
2 cubic yards of washed sand at 4/- = ...	0	8	0
3 cubic yards of broken stone at 2/6 = ...	0	7	6
6 cubic yards × 0·6 = 3·6 cubic yards =	£2	18	11

or 16/4 per cubic yard; therefore

	£	s.	d.
1 cubic yard of concrete = ...	0	16	4
Mixing and wheeling to site, per cubic yard =.......................................	0	2	6
Shuttering, per cubic yard = ...	0	1	0
Water pipes, &c., per cubic yard = ...	0	0	4
Total per cubic yard	£1	0	2

From the foregoing the following estimate can be formed :—

Foundations:

				£	s.	d.
Soft rock cutting...	3,400 cubic yards at 3/-			510	0	0
Hard rock cutting ...	1,980		7/6	742	10	0

Masonry:

				£	s.	d.
Masonry (building)	18,000		5/-	4,500	0	0
Stone	12,000		5/6	3,300	0	0
Mortar ...	6,000		24/-	7,200	0	0
Concrete...	2,340		20/2	2,359	10	0
Watering	18,000		-/2	150	0	0

Pointing:

				£	s.	d.
Inner face	2.200 square yards at 5/6			605	0	0
Outer face	2,150	,,	2/-	215	0	0
Superintendence, time-keeping, &c., say 5 per cent. of total.....................				979	2	0
Total cost				£20,561	2	0

Equal to £1 2s. 10d. per cubic yard of building.

The above requires, in addition, valve tower, discharge pipes, scour pipes, valves, railing, cleaning basin of reservoir, &c., &c., and is only given to indicate the cost per cubic yard with materials at about the above cost.

The Tausa Dam, Bombay, built 1886–92, cost for masonry only 14s. 8d. per cubic yard, but there were upwards of 407,407 cubic yards in the structure, which was built with hydraulic lime, the limestone being found near the site of the works, and labour was abundant and cheap.

The famous Furens Dam, built over 40 years ago, cost about 14s. per cubic yard, but such prices are exceedingly low, and depend largely on the essential material, cement—its cost on the site of the works influencing greatly the price per cubic yard.

The three principal items of cost in the foregoing estimate—viz., masonry building, stone, and mortar—come to 16s. 8d. per cubic yard.

APPENDIX II.

FORM OF SPECIFICATION.

THE forms of Specification and Schedule of Prices given in the following pages are subject to additions and alterations according to the customs prevailing, and the materials obtainable, in the district in which the dam may be constructed. There are customs, not to say prejudices, in various parts of the world which cannot be overlooked in arranging contracts, and what may be the general usage in England may not be satisfactory or applicable in other countries.

As an indication and outline the forms may be found useful, and, it is hoped, of assistance in the compilation of a specification that shall insure a satisfactory and honest piece of work being carried out—a guaranteed monument of stability and safety that may redound to the credit of the engineer and the contractor, whilst giving health and bounty to the district to be supplied with water.

It will be observed that no mention is made in the Specification of the gauge house and apparatus requisite for controlling the discharge, this not coming within the scope of the present treatise. Such information will be found in all works upon Hydraulic Engineering and Water Supply. Other additions may also be required, such as filter tanks, &c. The main object here is to confine attention to the design and construction of the dam itself and such preliminary details as form part of the work of the engineer.

[*Title of Dam to be specified here.*]

SPECIFICATION TO BE OBSERVED IN THE CONSTRUCTION OF THE WORKS NECESSARY FOR THE SUPPLY OF WATER TO   , &C., IN ACCORDANCE WITH THE DRAWINGS NUMBERED 1 TO   INCLUSIVE.

General Description.—The contractor is to provide the whole of the labour, carriage, tools, implements, tackle, legs, cranes, staging, pumps, and all other things requisite for the construction of the dam, valve tower, outlets, &c., to the forms and dimensions and in the position shown on the drawings.

Clearing Site of Reservoir.—The contractor shall remove all walls and fences, root out all trees, hedges, and brushwood, and clear away all mounds, loose earth, and rubbish within the area to be occupied by the reservoir. The material so obtained, as far as not claimed by the proprietors or tenants of the land, shall belong to the contractor.

Excavation. Foundations.—The loose and weathered portion of the rock to be stripped off to the solid for the whole width of the wall at its base and cut into horizontal beds. On the completion of the foundations of the wall a trench is to be cut in the solid rock without blasting, as shown on the drawings, and of such a depth as to effectually cut out all false seams, springs, veins of soft material, clay, &c., and to such further depths at any part of the foundations as may be found necessary by the nature of the ground, or as may be ordered by the engineer or his assistant. All loose and unsound places that may be caused during the execution of the works shall be excavated, and the foundations kept dry and clear of all obstructions, whether arising from springs or drains, &c. No extra allowance will be made on account of the depth of the stripping or trench until such depth exceeds ___ feet below the surface shown on the sections. For depths exceeding ___ feet the contractor will be paid according to the rates stated by him in the schedule for the various depths mentioned therein. The prices given by the contractor are to include the cost of close sheeting the trench if necessary, and keeping it free of water by means of such pumping engines and other appliances as may be necessary.

Before any concrete is put in the trench is to be inspected by the engineer or his assistant, to whom at least ___ days' notice must be given; and great care must be taken to have the bottom of the trench thoroughly cleaned out, scraped, and watered before putting in the concrete, so that a perfect and water-tight junction may be made. This applies equally to the whole foundation of the dam. Notwithstanding this inspection, however, the responsibility of making the reservoir water-tight shall rest with the contractor.

Waste.—The excavated waste from the stripping and trench shall be removed and deposited at such a distance and place from the works as may be indicated by the engineer or his assistant.

Concrete.—The cement concrete used for filling in the trench in the foundations shall be composed of ___ parts by measure of cement to ___ parts by measure of sand and ___ parts by measure of hard sharp stone, the sand and cement to be thoroughly well mixed before the water is added, the cement and sand to be worked into a mortar the consistency of which is such as will incorporate readily with the stone. The stone is to be perfectly free from sand or dirt and not larger than $2\frac{1}{2}$ inches cube or smaller than $1\frac{1}{2}$ inch cube, to be well wetted before being mixed with the cement mortar, and the whole aggregate to be well turned over to ensure perfect incorporation. The ingredients are to be mixed by machines of approved construction or by shovels upon a specially close-jointed timber floor of ample area and conveniently near the site of the works.

In no case is the concrete to be tipped or shot into the foundations, but to be laid in carefully in horizontal layers of one foot in thickness, to be afterwards rammed with wooden or iron rams until a film of water appears upon its surface, but not to make it quake.

No concrete or mortar is to be worked up the second time after partial setting, nor to be applied to a dry surface.

Should the weather be dry immediately following the completion of any portion of the work, the concrete, when sufficiently set, shall be regularly drenched with water in order to prevent the cracking of the surface.

Any portion that is left for a few days is, before placing more concrete upon it, to have its surface well swept, picked, and cleaned, so that a perfectly water-tight joint may be made.

The contractor will be obliged to remove and replace all concrete which in the opinion of the engineer or his assistant has been injured by frost, &c., while the works are in progress or during the period of maintenance.

Sand.—The whole of the sand used in the works, where not otherwise specified, is to be clean and sharp, not over coarse, and to be well washed before use.

Cement mortar.—The cement mortar is to be composed of parts by measure of cement to parts by measure of sand, to be thoroughly well mixed before applying the water, and each quantity of mortar made is to be brought to the same consistency as that previously used, sufficient only being made for the immediate requirements, all mortar showing signs of setting to be rejected.

The ingredients are to be mixed by machines of approved construction as in the case of the concrete.

Cement.—The Portland cement shall be of the best quality and from manufacturers of the highest standing, approved of by the engineer; it shall be of such a fineness as to pass a sieve of 14,400 holes per square inch, leaving a residue of not more than 10 per cent. Briquettes are to be made from the cement with not more than 20 per cent. of water by gauging, the same to stand after seven days' immersion in water 350 lbs. per square inch tensile strain, and after 28 days 500 lbs. per square inch, or an increase of not less than 25 per cent. between the two tests ; the strain to be applied at the rate of 400 lbs. per minute. The cement must be on the ground at least one month before it is required, to allow time for testing. The engineer or his assistant shall take the requisite samples out of each delivery (for the purpose of making such tests as he may desire), and should any delivery fail to stand the specified tests it shall be at once removed by the contractor and replaced by him with cement of the proper quality.

The cement when received on the works is to be well and dryly stored in a shed with raised floor built for the purpose by the contractor, space being provided for cooling. The cement to be stored in heaps of not more than six sacks depth, and not less than a week before using to be spread out, on the floor provided, to one foot depth, being turned every alternate day.

Proper measuring vessels are to be provided and constantly used by the contractor, each cubic foot holding not less than 90 lbs. of cement; if, owing to the fineness of the cement, not more than 85 lbs. can be contained in the cubic foot gauge box, the box is to be increased in size so as to hold 90 lbs. when striked, it being understood that this is the required weight per cubic foot gauge for the whole of the work.

Cement testing machine, &c.—A cement testing machine of approved make, together with the moulds, water tanks for briquettes, and other requisites shall be provided by the contractor for the use of the engineer or his assistant, and be at their disposal at any time when required, the same to be kept in a lock-fast house and the key in the possession of the engineer. On the completion of the works the contractor shall resume possession of the machine.

Stone and quarry.—The stone used in the work is to be of the best hard and sound class found upon the site of the works, free from seams and cracks, and to be approved of by the engineer. No quarrying operations are to be carried on within or below the overflow level of the reservoir or within   yards from the down stream face of the dam.

Construction.—The dam is to be constructed of uncoursed rubble masonry throughout, anything approaching regular horizontal joints to be carefully avoided, pains being taken to preserve a good bond throughout the whole breadth of the work; the whole to be carried up as much as possible of an uniform height, and in no case is there to be a greater difference of the levels in any parts of the same building than   feet without the special permission of the engineer.

All the building stones used are to be placed on their natural bed, and none are to be of less size than $\frac{1}{2}$ cubic foot, all weak corners being knocked off before placing in position, every stone to be laid full in mortar, each one being selected so as to roughly fit the place it is laid in and driven home with a light mallet, all spaces between it and the adjacent stone to be filled flush with mortar, spalls or small stones to be inserted in the mortar between the joints, care to be taken that the mortar and stone do not dry before setting the corresponding course or that any dry work or hollow spaces occur in the work.

The face stones and those used for the valve tower and gallery are to be specially selected, and, where necessary, cut to their true centres from templates. Grouting will in no case be permitted.

The face joints on the up stream face to be left with a 2 inch depth for pointing with a specially prepared cement mortar, consisting of 1 of cement to 1 of a finely ground hard sand by measure, to be slightly stiff for use, and driven in tightly into the joints with wooden implements. The down stream face and valve tower joints to be clean trimmed and pointed with the same strength of mortar as used in the construction.

The valve tower and connection to the outlet gallery to be built with a cement mortar of the same mixture as that for the pointing of the up stream face.

Waste weir.—The waste weir is to be constructed to the dimensions and in the manner shown on the drawings, the paved apron to be built with specially selected stone set in a cement mortar of 1 of cement to 2 of sand by measure and evenly pointed on the surface so as to offer the least friction to the flow of the water.

Protection of works.—Every precaution is to be taken to protect the works from the weather; and any work that may be injured by frost or otherwise is to be made good by the contractor.

Cast iron pipes.—The pipes required are to be cast vertically in dry sand moulds and to be of uniform bore and thickness throughout, to be of the best grey metal remelted from the cupola, and to be perfectly free from flaws and defects of any kind.

All pipes and special castings in which any imperfections shall appear, or wherein any sand holes or air holes shall be found plugged up, or shall not agree with the terms of the specification, will be rejected, and must be broken up in the presence of the engineer or his assistant.

The pipes are to be carefully coated on the inside only with coalpitch and oil, according to Dr. R. A. Smith's patent process, the coating to be applied at a proper heat and in a proper manner before any rust sets in.

The exterior of the pipes must be cleaned of all tar, oil, or grease, and be well picked with a sharp implement on the outer surface to roughen it, each pipe before being set in position to be struck with a heavy hammer, and should there be by this test any false ring, or doubt as to its soundness, it must be rejected and broken up. When set in position and bolted up the pipes are to be tested to a pressure of 250 lbs. per square inch by an approved machine. The pipes whilst this pressure is on them must be rapped with a hand hammer from end to end, so as to discover whether there are any sandy, porous, or blown places, any sign of leakage at the joints to be rectified before they are built in. The testing to be done by the contractor at his own expense and in the presence of the engineer or his assistant.

Bolts and nuts, &c.—The wrought iron in the bolts, nuts, &c. must be of the best make. The contractor is to state what class he proposes to use, which must bear a test of 24 tons to the square inch without breaking, and must be capable of being bent, whilst cold, to a right angle without fracture ; it must also bear a tensile strain of 10 tons to the square inch without permanent set. Any of the wrought iron bearing traces of oxidation will not be used in the work.

Valves.—The sluice valve for the scour pipe and the valves in the valve tower, &c., must be made of the best gun metal and of approved pattern. They must also be tested to 250 lbs. per square inch, and any leakage or defect must be immediately rectified or the valve rejected, it being understood that the valves shall be watertight and work satisfactorily.

Building in of pipes.—Great care is to be exercised in the building in of the scour pipes and the connections of the discharge pipes from the inner to the outer part of the valve tower that a secure joint is made between the rubble masonry, &c., and the iron pipes.

General Conditions.

Omissions of details on plans, &c.—All the works, although parts of the same only may be marked on the plans and sections, are part of the contract, and included therein, as much as if such works had been particularly set forth and described in the specification also. Such of the works as may be mentioned in the specification only, without being drawn on the plans or sections, are included in the contract as if they had been particularly drawn on the plans and sections also, and as there may be details not particularly mentioned in the specification, nor drawn in the plans and sections, the contract must be taken to include all such details as may have thus been omitted, it being clearly understood that the contractor is to execute all the works requisite for the perfect completion of each and every of the several parts according to the true intent and meaning of the specification.

Drawings.—Wherever the dimensions are marked on the plans and sections, or described in the specification, such dimensions are to be considered as correct, although not exactly corresponding with the measurements from the scales, which must be referred to only when the dimensions are not so marked or described. The drawings to larger scales, and those showing particular parts of the work, must be taken as more correct than those to smaller scales, or for more general purposes. The contractor must, without any extra charge, execute the several parts of the work in strict compliance with the working or detailed drawings that may from time to time be furnished by the engineer.

Custody of drawings.—The drawings referred to in the specification, and signed by the contractor, shall remain in the custody of the engineer during the progress of the works, the contractor being allowed access to them at all reasonable times, for the purpose of taking copies of the whole or any portion of them he may require.

Alterations and deviations. — The engineer may increase, diminish, or alter the work without vitiating the contract. The value of such increases or diminutions or alterations being ascertained from the rates and prices stated in the schedule annexed to the contract, and being added to or deducted from the contract sum as the case may be.

Extra work.—No extra work of any kind or description whatever will be paid for by   unless it has been executed under a written order signed by the engineer or his assistant, and a weekly bill of all such work, or such parts thereof as may be executed, must be delivered by the contractor to the engineer or his assistant during the following week, and the non-delivery of such bill at the proper and stated time will be considered as an abandonment on the part of the contractor of any claim for the amount of such work, and as exonerating the from all liability relating thereto. The value of all such extra work will be paid for according to the rates and prices stated in the schedule annexed to the contract.

Measurements.—All measurements must be the net dimensions of the work when finished, notwithstanding any custom that may prevail to the contrary. All metal used in the work must be weighed, and a note of the weight delivered to the engineer or his assistant previously to the work being fixed, the weighing being done in such a manner as will allow of the weight of any particular part of the work being ascertained.

Materials and workmanship.—The whole of the works, both as regards quality of materials and mode of execution, must be performed and completed in the most approved, workmanlike, and substantial manner, under the direction, and to the entire satisfaction, of the engineer.

Power to inspect materials.—The engineer or his assistant may at any time inspect the materials and manufacture of the various parts of the works, and if the contractor refuse to allow such inspection the work will be deemed insufficient and not in accordance with the terms of the specification.

Use of land.—The contractor shall not use the land forming the site of or connected with the works for any other purpose whatever than the proper carrying on of the works.

Temporary huts. — Should the contractor require to erect temporary huts for his workmen, or other temporary buildings, the sites must be approved by the proprietors of the land and by the engineer or his assistant.

Trespassing.—The workmen are to be prohibited from trespassing on the neighbouring lands, from disturbing the sheep or cattle, and from killing, disturbing, or otherwise injuring the fish or game. The contractor is to dismiss at once, on the application of the engineer or his assistant, any workman or foreman so engaged, and to punish, so far as he has the power, all trespassers and poachers.

Neither the contractor nor any one in his employment will be allowed to keep a dog on the works, nor to fish in any of the streams.

Accidents.—The contractor must use all reasonable precautions to prevent accidents in the carrying out of the works ; he shall provide proper storage for explosives, and careful and experienced men for handling them, and he must undertake liability for all accidents or damage that may occur in connection with the carrying out of the works, whether to his own workpeople, to the general public, or to property of any kind ; and must free and relieve the the engineer, or his assistant, from any claims that may be made against them in respect to such accidents or damage.

In case of dispute between him and any party claiming compensation, or if the claim is made directly against the , the amount of such compensation shall, with the consent of the party claiming compensation, be fixed by the engineer, whose decision on that point shall be final, the amount to be deducted from any sum that may be due or may become due to the contractor.

The contractor must also use every precaution to guard against accidents or injury to the works by reason of floods, or by the action and pressure of water, or by frosts, slips, or leakage, and should any such damage take place from such or other causes he shall forthwith repair and make good the same at his own expense.

Removal of rubbish.—The contractor shall from time to time remove all surplus and objectional materials, waste, or rubbish from the works, and from any land or premises where any portion of the work may be carried on.

Foreman or workmen.—The contractor shall employ at his own cost and charge a competent foreman or engineer, who is to be constantly on the work, to ensure efficient control and superintendence, and who shall be duly authorised to act and receive instructions from the engineer or his assistant, and any instructions given shall have equal validity as if given to the contractor himself. No person shall be employed or allowed to remain on the work or any part thereof who shall be objectionable to the engineer.

Contractor to attend to engineer's orders and directions.—The contractor shall attend to and execute without delay all orders and directions which may from time to time be given by the engineer or his assistant in connection with the contract, and if he refuse to comply with all such orders and directions, or become bankrupt, or insolvent, or does not proceed with all due diligence and expedition within twenty-four hours after a written notice requiring the same has been delivered to him or his foreman, the engineer may use, free of cost and charge for wear and tear, all or any of the contractor's men, tools, implements, and materials which may be on the work or in use at the time, and also employ other men, tools, implements, and materials to perform such works as he may require and direct, agreeably with the specification ; and all the costs, charges, and expenses of the same shall be deducted from any amount that may be due to the contractor and retained by the   in re-imbursement of all such costs, charges, and expenses.

Notice to the contractor.—All notices to the contractor shall be given in writing, and the delivery of them to his foreman, or at any of the contractor's usual places of business or residence, shall be deemed sufficient service.

Imperfections or insufficient workmanship.—If at any time

during the progress of the works, or within twelve calendar months after the completion, any imperfections, leakage, or insufficient workmanship shall appear, the contractor shall forthwith make good the same at his own expense; the true intent and meaning of the specification being that the whole of the works shall be delivered up to the   properly and completely finished and perfect in all their parts, and in conformity in every respect with the contract.

Watchman.—From sunset to sunrise, and when required, a watchman is to be kept on Sundays and all other times when the works are not in progress until they are completely finished.

Spirituous liquors.—The contractor shall not sell or allow to be sold or brought within the limits of his work any spirituous liquors, and will in every way discountenance their use by persons in his employ.

Inspector of works or clerk of works.—The engineer shall appoint a clerk of works or inspector to take charge of and supervise the various sections of the work during its construction, and the said assistant of the engineer shall have it in his power to dismiss any foreman or workman who may be unskilled or inefficient, or who may refuse or neglect to attend to their orders or instructions, or to those of the engineer given through him.

Office for assistant.—The contractor shall provide an office for the clerk of works or inspector appointed, the same to be provided with stove, desk with drawer, lock and key, the floor to be boarded, and windows and doors provided where necessary with necessary fastenings.

Maintenance of works.—The contractor, notwithstanding the use of the said works for the purpose of water supply and storage, shall be responsible for, and shall maintain and uphold in a sound and perfectly water-tight condition, every part of the works for a period of twelve months from the date of the engineer's certificate of completion. In the event of the contractor failing to make any necessary repairs when called upon to do so the same may be done by the engineer or his servants and the cost thereof retained from any sum that may be due or become due to the contractor.

Period of completion.—The entire work shall be warranted by the contractor and completed within   from the date of the written order to proceed with the work, under a forfeiture of   per week, to be paid to and retained by   , by way of liquidated and ascertained damages, and not by way of penalty, for each week and every week the work shall remain unfinished after the expiration of the period above mentioned.

Completion of works.—The works shall not be deemed finished or complete until they shall have been certified to be so in writing by the engineer.

Power to delay works.—The engineer may delay the progress of the works without vitiating the contract, and grant such extension of the time for the completion of the contract as he may think proper and sufficient in consequence of such delay, and the contractor shall not make any claim for compensation or damage in relation thereto.

Disputes.—If at any time during the progress or after the completion of the works any disputes or differences shall arise as to the manner of executing the works, or as to the quality of the materials employed, or as to any matter of charge or account between the   and the contractor, or as to any other matter or thing connected with the contract, they shall be referred to and finally settled by   as sole arbiter, whose decision shall be final and binding on both parties.

Payments.—Payments shall be made upon the recommendation of the engineer, the first of such payments whenever he may consider that   per cent. of the work contracted for has been executed, and the subsequent payments from time to time when it shall appear to him that a similar proportion of the work contracted for has been delivered and executed since the preceding payment. All such payments will be at the rate of 80 per cent. of the works so considered to have been executed, whether contract or extra works. One half of the balance due will be paid on the completion of the works, and the remainder at the expiration of twelve calendar months after their completion, under deduction of all claims against the contractor.

Contract.—Before commencing the works the contractor shall enter into a formal contract with the   for the execution of the work, providing such security as may be satisfactory to them, which contract shall contain all the usual provisions and stipulations, and he shall pay one half of the cost of preparing the same.

Tenders.—Tenders on the printed form with schedule must be filled up and addressed to the   , and endorsed   , and are to be delivered at   on or before   o'clock in the morning of   inst. The amount of the tender is to be such as to cover all contingencies and omissions in the drawings and specification, the prices in the schedule being for the purposes of computing the value of extra work.

The   do not bind themselves to accept the lowest or any tender.

Date   (Signed)

SCHEDULE OF PRICES TO ACCOMPANY TENDER.

Excavator's Work.

Excavating and removing from
    site of works when the depth
    does not exceed 6 feet......... at   per cubic yard.
Ditto   ditto   12 ,, ... ,,   ,,   ,,
Ditto   ditto   18 ,, ... ,,
Ditto   ditto   24 ,, ... ,,
Ditto   ditto   30 ,, ... ,,
Ditto   ditto   36 ,, ... ,,
Ditto   ditto   40 ,, ... ,,
Clearing ground and removing
    rubbish within interior of
    reservoir .... ,,   ,,   ,,
NOTE.—All the above prices are also to include all
    matters and things referred to in the specification
    as to maintaining the excavation open and dealing
    with water, &c.

Mason's Work.

Concrete .... at   per cubic yard.
Stone................................. ,,   ,,   foot.
Ditto in mortar.................... ,,   ,,
Cement   ,, ... ,,   ,,   yard.
Cut and tooled stone for valve
    tower, arch, &c. .... ,,   ,,   foot.
Ditto   ditto   ditto in mortar... ,,   ..

Ironwork, &c.

Wrought iron rolled beams...... at   per cwt.
Ditto T irons and L irons ...... ,,   ,,
Ditto bolts and nuts............... ,,   per lb.
Cast iron pipes with special
     flanges .... ,,   per cwt.
Ditto   ditto ordinary flanges ,,   ,,
Bronze valves .... ,,   per lb.
Railing .... ,,   per lineal yard.
Painting .... ,,   per superfic. yard.

Labour only.

Excavator .... at   per hour.
Labourer to ditto .... ,,   ,,
Mason .... ,,
Labourer to ditto .... ,,
Horse, cart, and man .... ,,
Extra horse .... ,,
Fitter .... ,,
Rivetter .... ,,

Signature

Address

Date

FORM OF TENDER.

Proposal of   *for Building Masonry Dam, &c.*

Date

Messrs. The

GENTLEMEN,—   offer to execute the whole of the works, to furnish all the materials, labour, plant, complete and maintain the work in accordance with the plans, sections, and specification, and with such direction as may from time to time be received from the engineer, for the sum of sterling.

The above amount includes all materials, workmanship, labour, scaffolding, tools, and machinery, and every expense necessary for the completion of the whole work.

In the schedule   have given the prices for the various classes of work upon which all extra work will be calculated, and   undertake to complete the whole contract within   months from the date of receiving order to proceed.

Gentlemen,

Your obedient servant,

(Signed)

Amount of tender   Address

References

Address

APPENDIX III.

PURITY OF WATER.

AN important part of the investigation required previously to the erection of a dam is the question of the purity of the water which is to be retained by the dam from the drainage area, and any one in search of a pure supply of water will be glad of a few simple tests and instructions which may enable him, without the necessity of an expert's assistance in the preliminary work, to ascertain with a certain degree of probability whether the water which it is proposed to utilize is of sufficient purity to warrant its recommendation for domestic supply. Any water that it might be proposed to impound for the supply of a large town would, undoubtedly, require to be subjected to a searching examination before expense was incurred for the works necessary for a constant supply, and the following extract* is only given here as a useful guide to the prospector for water :—

* From the " American Engineering and Mining Journal."

TEST FOR PURITY OF DRINKING WATER.

By FRANCIS WYATT, Ph.D.

When chemists apply the word " pure " to water they of course only do so in a comparative sense, because perfectly pure water does not exist in nature. Even in its primary form of rain it contains some traces of ammonia and nitrates, derived from the atmosphere, and it always becomes more or less charged with earthy and saline matters before it reaches the streams. The rivers are charged with impurity and refuse from towns on their banks, and the water becomes gradually more dangerous ; and, although it is somewhat purified by oxidation and the absorbent action of vegetation, it requires the most conscientious and watchful care in the reservoirs of great communities.

All upland surface waters vary in quality in accordance with the nature of their surrounding conditions, but they are characterised as " pure " and accepted as satisfying all necessary conditions for drinking and household purposes when they have no disagreeable taste or smell, when they are only of medium hardness and are free from excess of salt, and when they have no poisonous minerals and only a minimum of organic contamination.

In order to ascertain whether or not a given source of water really fulfils the needed requirements it must be subjected to the closest scientific scrutiny, for little reliance can be placed upon the public taste. Nothing less than the determination, within reasonable limits of accuracy, of the amount of matter in the water, and of the probable origin of the impurities, is of much public value, and this determination is only

M.D.   K rendered possible by accurate chemical and biological examinations. It is no exaggeration to say that, from a sanitary standpoint, the difficulties in the way of assigning proper importance to the various ingredients discovered in an analysis are well-nigh insurmountable. In fact, so great are they that careful chemists invariably make their reports and conclusions only after comparing their own conditions and results with the results recorded for similar conditions over a long period of years by established authorities.

The main causes of perplexity and doubt are not the inorganic salts, which all drinking waters contain in more or less abundance, but those complex and enigmatical bodies which have come to be classified under the heads of organized and unorganized organic matter. The real difficulty in the sanitary analysis of a water is to show—first, how far it is contaminated with bacteria or micro-organisms; and, second, to what extent it is capable of affording nutrition to such organisms in the form of readily decomposable feeding material.

There being no prescribed rule of general applicability by which the interpretation of a purely chemical analysis of water can be made by the ordinary reader, the mere publication of certain analytical results, even by the best authorities, are practically without significance, since they are commonly unaccompanied by any detailed and intelligible or popular explanations.

The following figures approximate to what I regard, after a wide experience, as the typical salient points in a perfectly safe surface water ; 100,000 parts of a potable water may contain : total inorganic solid residue, 5 up to 50 parts ; chlorine, 0·10 up to 1 part; phosphates, none ; nitrites, none ; free ammonia, 0·005 parts; albuminoid ammonia, 0·015 parts; oxygen consumed, 0·250 parts ; poisonous minerals, none. It is possible that such an analysis as this would not call forth very extensive comment, but in cases which show any notable increase in any or all of my figures, and where some turbidity, or colour, or taste, or smell, are also noted, the report of the analyst should be supplemented by some remarks upon, (1) the behaviour on ignition of the total solid residue, (2) the quantity of the free and albuminoid ammonia and their proportionate relation to each other, (3) the quantity of oxygen absorbed by a normal sample of the water from permanganate of potash, and the time required for such absorption, (4) the amount of nitrous acid or nitrites, (5) the quantity of chlorine. These annotations should be further accompanied by explanatory foot-notes, stating in a general way, for the benefit of the uninitiated, that when residual solids turn very black on ignition, and emit an odour resembling burnt hair, they point to the probable presence of animal matter. That when the quantity of free ammonia is very high, and especially when it is accompanied either by nitrous or phosphoric acid, or by both, the indications are strongly in favour of recent sewage contamination; that when, in addition to much free ammonia, the albuminoid ammonia exceeds the figures 0·015, it is to be regarded as a measure of the potential putrescible animal matter which still exists in the water, and which will afford food for innumerable ferments of all species. That, when the quantity of " oxygen absorbed," under certain conditions (by Forschammer's test), is more than 0·250, it is to be regarded as entirely confirmatory of the other data, as is the amount of chlorine when, in waters far from the sea, or from salt-bearing strata, it exceeds a maximum of two parts per 100,000.

Such information as this, modified to suit varying circumstances, would be of material assistance, even though it be based upon purely chemical tests. It might even become a generally applicable rule, and could be regarded as entirely conclusive if the figures obtained, and considered as a whole, were sufficiently abnormal to be startling.

Such cases, however, seldom occur in actual practice, the

numbers rarely exhibit any positive regularity, and usually only one or two of them are sufficiently high to excite suspicion. Here, therefore, the rule would no longer apply, and here commences the perplexity, which calls for additional evidence through the medium of the microscope. To again put it more plainly, the vitality of the micro-organisms present in the water must be made the gauge of its potential organic impurity in all cases of serious doubt.

During the past few years there has been immense progress in biological science, but it is not yet sufficiently great to enable us to say with authority, off hand, whether the organisms found in a water supply are disease-producing or innocuous. It does enable us, however, to determine the extent of their vitality and their approximate numbers, and from this we may deduce the value of the water as a nutrient medium for septic or other dangerous microbes that might gain access to it. The method most commonly used for the bacteriological examination of water consists in mixing a measured quantity of it (1 c.c.) in a test tube, with sterilized nutrient gelatin, and pouring the mixture upon a sterilized glass plate. After the gelatin has solidified the glass plate is placed in a damp chamber and kept at a uniform temperature of about 70° to 80° Fahr. for five or six days. Colonies of microbes have by this time made their appearance, and may be counted with the microscope by the aid of Wolffhugel's apparatus.

There can be no doubt of the excellence of this method, but in my own practice I have found it too tedious for commercial work, and have replaced it, after many experiments, by one which is more simple and expeditious, and gives equally satisfactory information. As culture media I use neutral, slightly acid, and slightly alkaline decoctions of malt extract sterilized by boiling in Pasteur flasks. These flasks are always kept on hand ready for immediate use, and the sterilization having destroyed all forms of life, and the trace of outside germs being absolutely prevented, the liquids preserve themselves indefinitely without losing their brilliancy or undergoing the slightest change. When a water is submitted for analysis the first step is to shake it up well, to insure perfect homogeneity. When this has been done 1 c.c. of it is introduced into each of the three flasks by means of a sterilized pipette. The flasks are then kept side by side at a uniform temperature of 80° Fahr. for 48 hours, and the behaviour of each flask is very carefully watched, and a note is made from time to time of the fermentative activity of each. The operation is always conducted on exactly the same lines, and yields data which, when compared with that resulting from the chemical tests, is most useful and reliable.

For example, if a water which has yielded a marked excess of "albuminoid ammonia" in the chemical tests quickly clouds up and develops very active fermentation in the flasks, we have immediate and undoubted proof of the existence of large numbers and various species of bacteria, and since their activity has been developed in such widely differing nutritive media it is safe to regard them as probably containing many forms of a highly objectionable kind, and to denounce the water as impure and unfit for drinking purposes.

If it were argued against this condemnation that nearly all surface waters are more or less contaminated with vegetable matter, and that this matter is always accompanied by infusorial and other lower forms of vegetable life which grow in nutrient solutions, and which are nevertheless assumed to be harmless, it may be answered that this is a question for the pathologist.

The province of the chemist is to determine the condition of a water, and what I would emphasise is that the presence of very active bacteria in large quantities growing in neutral, acid, and alkaline media, under the circumstances named, would, to my mind, invariably indicate sewage or animal contamination, and would constitute a danger signal of the highest significance. There are circumstances which, in certain cases, make it extremely difficult, if not impossible, to obtain complete scientific examinations of a water supply, but I do not consider that this need necessarily preclude the adoption of rational precautionary measures. We owe the adoption of such measures as we may command, not only to ourselves, but to those about us, and the duty may be discharged by means of a few simple tests which are very easily performed.

The following tests require no apparatus that is not found in every household, and I believe they are sufficient to determine whether any given water of unknown quality is safe for drinking purposes :—

1. Pour a glass of the water into a decanter, cork it, and shake it up violently for a minute or two. If it develops any bad smell after the operation the water may be suspected of sewage or other animal contamination.

2. Add to a small glassful of the water two or three drops of diluted sulphuric acid and stir, then pour in about two drops of a weak solution of permanganate of potassium, or sufficient to colour it a faint rose. Cover the glass with a saucer, and leave it standing for ten minutes, when, if the rose colour has entirely disappeared, the water is probably unwholesome, and requires to be investigated.

3. Take a very clean, dry glass, and put into it a few drops of a solution of nitrate of silver, and then pour in a couple of ounces of the water. If it becomes milky, add to it a few drops of diluted nitric acid. If the milkiness does not now nearly all clear away the water will be proved to contain much chlorine, and, unless it be taken from some source near the coast or near to salt springs, is probably contaminated with sewage.

4. Take two white eight-ounce bottles with well fitting stoppers, and wash them thoroughly clean. Nearly fill one of them with the natural water, and the other with the water after boiling it for thirty minutes. Now put into each bottle a teaspoonful of pure white granulated sugar. Shake them until the sugar dissolves, and then place them side by side at a temperature of about 80° Fahr., and let them stand for three days. If the unboiled water rapidly clouds up, and shows a marked fermentation, emitting an odour faintly recalling rancid butter, it probably contains phosphates, and may be suspected of contamination with sewage. If the boiled water shows any signs of decomposition the suspicion of serious contamination will be confirmed.

5. Pour a small quantity of the water into a white saucer, and carefully add to it one drop of sulphuret of ammonia. If a dark colour is formed, which immediately disappears on the addition of one or two drops of pure hydrochloric acid, iron salts are present. If the dark colour does not disappear the water contains other and probably poisonous metals, and should at once be rejected.

If any or all of the first four of these rough tests be doubtful or unfavourable in their results they will render service of incalculable value, if only by pointing to the necessity for purifying the water before it can be drunk with safety.

How this purification should be performed is a matter upon which there are numerous suggestions and much difference of opinion; the best and most efficient manner of doing it in the household is by boiling it for thirty minutes. Dr. Miguel, of Paris, has dispelled all doubt upon this subject by publishing the following figures :—Bacteria in each c.c. of water, at 60° Fahr., 460,800; 122° Fahr., for ten minutes, 600; 140° Fahr., for ten minutes, 90; 158° Fahr., for ten minutes, 89; 176° Fahr., for ten minutes, 63; 195° Fahr., for ten minutes, 27; 212° Fahr., for ten minutes, none. This evidence is too striking to require any comment, and I will merely say in conclusion that if, after it is boiled, the water be first cooled, and then passed through an ordinary charcoal or other approved filter, its flat and uninviting taste will be immediately removed.

INDEX.

THE END.

BRADBURY, AGNEW, & CO. LD., PRINTERS, LONDON AND TONBRIDGE.

THE

WATER SUPPLY OF TOWNS

AND THE

CONSTRUCTION OF WATERWORKS:

*A PRACTICAL TREATISE FOR THE USE OF ENGINEERS
AND STUDENTS OF ENGINEERING.*

BY

W. K. BURTON, Assoc. Memb. Inst.C.E.

PROFESSOR OF SANITARY ENGINEERING IN THE IMPERIAL UNIVERSITY, TOKYO, JAPAN
CONSULTING ENGINEER TO THE TOKYO WATERWORKS
ENGINEER TO THE SANITARY BUREAU, HOME DEPARTMENT, JAPAN.

TO WHICH IS APPENDED

A PAPER ON THE EFFECTS OF EARTHQUAKES ON WATERWORKS,

By PROFESSOR JOHN MILNE, F.R.S.

With numerous Plates and other Illustrations.

OPINIONS OF THE PRESS.

"A thoroughly practical treatise, . . . which will unquestionably be of great use to water engineers."—*Local Government Journal.*

"It would be impossible within reasonable compass to afford anything like an adequate review of this volume. . . . The chapters dealing with the different qualities of waters, the quantity of water to be provided and its purification, settling reservoirs, filtration, distribution systems, prevention of waste of water, pipes for waterworks, provisions for the extinction of fire, have struck us as very good. The illustrations are excellent."—*Lancet.*

"We congratulate the author upon the practical common sense shown in the preparation of this work. . . . The plates and diagrams have evidently been prepared with great care, and cannot fail to be of great assistance to the student."—*Builder.*

"The whole art of waterworks construction is dealt with in a clear and comprehensive fashion."—*Saturday Review.*

London: CROSBY LOCKWOOD AND SON,

7, STATIONERS' HALL COURT, LUDGATE HILL, E.C.

BOOKS FOR ENGINEERS OF WATERWORKS.

THE WATER SUPPLY OF CITIES AND TOWNS (A Comprehensive Treatise on). By WILLIAM HUMBER, A.-M.Inst.C.E.; M. Inst. M.E Illustrated with 50 Double Plates, 1 Single Plate, Coloured Frontispiece, and upwards of 250 Woodcuts, and containing 400 pages of Text. Imp. 4to, £6 6s. elegantly and substantially half-bound in morocco.

LIST OF CONTENTS.

I. HISTORICAL SKETCH OF SOME OF THE MEANS THAT HAVE BEEN ADOPTED FOR THE SUPPLY OF WATER TO CITIES AND TOWNS. —II. WATER AND THE FOREIGN MATTER USUALLY ASSOCIATED WITH IT.—III. RAINFALL AND EVAPORATION.—IV. SPRINGS AND THE WATER-BEARING FORMATIONS OF VARIOUS DISTRICTS.—V. MEASUREMENT AND ESTIMATION OF THE FLOW OF WATER. —VI. ON THE SELECTION OF THE SOURCE OF SUPPLY.—VII. WELLS.—VIII. RESERVOIRS.—IX. THE PURIFICATION OF WATER. —X. PUMPS.—XI. PUMPING MACHINERY. —XII. CONDUITS.—XIII. DISTRIBUTION OF WATER.—XIV. METERS, SERVICE PIPES, AND HOUSE FITTINGS.—XV. THE LAW AND ECONOMY OF WATERWORKS.—XVI. CONSTANT AND INTERMITTENT SUPPLY. — XVII. DESCRIPTION OF PLATES.—APPENDICES, GIVING TABLES OF RATES OF SUPPLY, VELOCITIES, &C., &C., TOGETHER WITH SPECIFICATIONS OF SEVERAL WORKS ILLUSTRATED, AMONG WHICH WILL BE FOUND: ABERDEEN, BIDEFORD, CANTERBURY, DUNDEE, HALIFAX, LAMBETH, ROTHERHAM, DUBLIN, AND OTHERS.

RURAL WATER SUPPLY: A Practical Handbook on the Supply of Water and Construction of Waterworks for small Country Districts. By ALLAN GREENWELL, A.M.I.C.E., and W. T. CURRY, A.M.I.C.E. With Illustrations. Crown 8vo, 5s. cloth.

HYDRAULIC TABLES, CO-EFFICIENTS, AND FORMULÆ for Finding the Discharge of Water from Orifices, Notches, Weirs, Pipes, and Rivers. With New Formulæ, Tables, and General Information on Rainfall, Catchment Basins, Drainage, Sewerage, Water Supply for Towns, and Mill Power. By JOHN NEVILLE, C.E., M.R.I.A. Third Edition. Crown 8vo, 14s. cloth.

HYDRAULIC MANUAL. Consisting of Working Tables and Explanatory Text. Intended as a Guide in Hydraulic Calculations and Field Operations. By LOWIS D'A. JACKSON. Fourth Edition, Enlarged. Large crown 8vo, 16s. cloth.

WATER ENGINEERING: A Practical Treatise on the Measurement, Storage, Conveyance, and Utilisation of Water for the Supply of Towns, for Mill Power, and for other Purposes. By CHARLES SLAGG, A.-M Inst.C E. Second Edition. Crown 8vo, 7s. 6d. cloth.

DRAINAGE OF LANDS, TOWNS, AND BUILDINGS. By G. D. DEMPSEY, C E. Revised, with large Additions on RECENT PRACTICE IN DRAINAGE ENGINEERING, by D. KINNEAR CLARK, M.Inst.C.E. Third Edition. Fcap. 8vo, 4s. 6d. cloth.

RIVER BARS: The Causes of their Formation, and their Treatment by "Induced Tidal Scour"; with a description of the Successful Reduction by this Method of the Bar at Dublin. By I. J. MANN, Asst. Eng to the Dublin Port and Docks Board. Royal 8vo, 7s. 6d. cloth.

. For NOTICES OF THE PRESS *concerning the above Works, see Catalogue at end of this Volume, pp. 10 and 11.*

LONDON: CROSBY LOCKWOOD AND SON,

7, STATIONERS' HALL COURT, LUDGATE HILL, E.C.

A

CATALOGUE OF BOOKS

INCLUDING NEW AND STANDARD WORKS IN

ENGINEERING: CIVIL, MECHANICAL AND MARINE;
ELECTRICITY AND ELECTRICAL ENGINEERING;
MINING, METALLURGY; ARCHITECTURE,
BUILDING, INDUSTRIAL AND DECORATIVE ARTS;
SCIENCE, TRADE AND MANUFACTURES;
AGRICULTURE, FARMING, GARDENING;
AUCTIONEERING, VALUING AND ESTATE AGENCY;
LAW AND MISCELLANEOUS.

PUBLISHED BY

CROSBY LOCKWOOD & SON.

MECHANICAL ENGINEERING, etc.

D. K. Clark's Pocket-Book for Mechanical Engineers.

THE MECHANICAL ENGINEER'S POCKET-BOOK of Tables, Formulæ, Rules, and Data: A Handy Book of Reference for Daily Use in Engineering Practice. By D. KINNEAR CLARK, M. Inst. C.E., Author of "Railway Machinery," "Tramways," &c. Third Edition, Revised. Small 8vo, **700 pages, 6s.** bound in flexible leather cover, rounded corners.

SUMMARY OF CONTENTS.

MATHEMATICAL TABLES.—MEASUREMENT OF SURFACES AND SOLIDS.—ENGLISH WEIGHTS AND MEASURES.—FRENCH METRIC WEIGHTS AND MEASURES.—FOREIGN WEIGHTS AND MEASURES.— MONEYS.—SPECIFIC GRAVITY, WEIGHT AND VOLUME.—MANUFACTURED METALS.—STEEL PIPES.— BOLTS AND NUTS.—SUNDRY ARTICLES IN WROUGHT AND CAST IRON, COPPER, BRASS, LEAD, TIN, ZINC.—STRENGTH OF MATERIALS.—STRENGTH OF TIMBER.—STRENGTH OF CAST IRON.—STRENGTH OF WROUGHT IRON.—STRENGTH OF STEEL.—TENSILE STRENGTH OF COPPER, LEAD, ETC.—RESIST. ANCE OF STONES AND OTHER BUILDING MATERIALS.—RIVETED JOINTS IN BOILER PLATES.—BOILER SHELLS.—WIRE ROPES AND HEMP ROPES.—CHAINS AND CHAIN CABLES.—FRAMING.—HARDNESS OF METALS, ALLOYS AND STONES.—LABOUR OF ANIMALS.—MECHANICAL PRINCIPLES.—GRAVITY AND FALL OF BODIES.—ACCELERATING AND RETARDING FORCES.—MILL GEARING, SHAFTING, &C.—TRANSMISSION OF MOTIVE POWER.—HEAT.—COMBUSTION: FUELS.—WARMING, VENTILATION, COOKING STOVES.— STEAM.—STEAM ENGINES AND BOILERS.—RAILWAYS.—TRAMWAYS.—STEAM SH.PS.—PUMPING STEAM ENGINES AND PUMPS.—COAL GAS, GAS ENGINES, &C.—AIR IN MOTION.—COMPRESSED AIR.—HOT AIR ENGINES.—WATER POWER.—SPEED OF CUTTING TOOLS.—COLOURS.—ELECTRICAL ENGINEERING.

₊ OPINIONS OF THE PRESS.

"Mr. Clark manifests what is an innate perception of what is likely to be useful in a pocket-book, and he is really unrivalled in the art of condensation. Very requently we find the infcrmation on a given subject is supplied by giving a summary description of an experiment, and a statement of the results obtained. There is a very excellent steam table, occupying five-and-a-half pages; and there are rules given for several calculations, which rules cannot be found in other pocket-books, as, for example, that on page 497, for getting at the quantity of water in the shape of priming in any known weight of steam. It is very difficult to hit upon any mechanical engineering subject concerning which this work supplies no information, and the excellent index at the end adds to its utility. In one word, it is an exceedingly handy and efficient tool, possessed of which the engineer will be saved many a wearisome calculation, or yet more wearisome hunt through various text-books and treatises, and, as such, we can heartily recommend it to our readers, who must not run away with the idea that Mr. Clark's Pocket-book is only Molesworth in another form. On the contrary, each contains what is not to be found in the other; and Mr Clark takes more room and deals at more length with many subjects than Molesworth possibly could."—*The Engineer.*

"It would be found difficult to compress more matter within a similar compass, or produce a book of 650 pages which should be more compact or convenient for pocket reference. . . . Will be appreciated by mechanical engineers of all classes."—*Practical Engineer.*

"Just the kind of work that practical men require to have near to them."—*English Mechanic.*

A

MR. HUTTON'S PRACTICAL HANDBOOKS.

Handbook for Works' Managers.

THE WORKS' MANAGER'S HANDBOOK of Modern Rules, Tables, and Data.

For Engineers, Millwrights, and Boiler Makers; Tool Makers, Machinists, and Metal Workers; Iron and Brass Founders, &c. By W. S. HUTTON, Civil and Mechanical Engineer, Author of "The Practical Engineer's Handbook." Fifth Edition, carefully Revised, with Additions. In One handsome Volume, medium 8vo, price 15s strongly bound.

☞ *The Author hav'ng compiled Rules and Data for his own use in a great variety of modern engineering work, and having found his notes extremely useful, decided to publish them—revised to date—believing that a practical work, suited to the* DAILY RE-QUIREMENTS OF MODERN ENGINEERS, *would be favourably received.*

*** OPINIONS OF THE PRESS.

"Of this edition we may repeat the appreciative remarks we made upon the first and third. Since the appearance of the latter very considerable modifications have been made, although the total number of pages remains almost the same. It is a very useful collection of rules, tables, and workshop and drawing office data."—*The Engineer*, May 10, 1895.

"The author treats every subject from the point of view of one who has collected workshop notes for application in workshop practice, rather than from the theoretical or literary aspect. The volume contains a great deal of that kind of information which is gained only by practical experience, and is seldom written in books."—*The Engineer*, June 5, 1885.

"The volume is an exceedingly useful one, brimful with engineers' notes, memoranda, and rules, and well worthy of being on every mechanical engineer's bookshelf."—*Mechanical World*.

"The information is precisely that likely to be required in practice. . . . The work forms a desirable addition to the library not only of the works' manager, but of anyone connected with general engineering."—*Mining Journal*.

"Brimful of useful information, stated in a concise form, Mr. Hutton's books have met a pressing want among engineers. The book must prove extremely useful to every practical man possessing a copy."—*Practical Engineer*.

New Manual for Practical Engineers.

THE PRACTICAL ENGINEER'S HANDBOOK, Comprising a Treatise on Modern Engines and Boilers, Marine, Locomotive, and Stationary.

And containing a large collection of Rules and Practical Data relating to recent Practice in Designing and Constructing all kinds of Engines, Boilers, and other Engineering work. The whole constituting a comprehensive Key to the Board of Trade and other Examinations for Certificates of Competency in Modern Mechanical Engineering. By WALTER S. HUTTON, Civil and Mechanical Engineer, Author of "The Works' Manager's Handbook for Engineers," &c. With upwards of 370 Illustrations Fifth Edition, Revised, with Additions. Medium 8vo, nearly 5co pp., price 18s. strongly bound. [*Just published.*

☞ *This work is designed as a companion to the Author's "WORKS' MANAGER'S HANDBOOK." It possesses many new and original features, and contains, like its predecessor, a quantity of matter not originally intended for publication, but collected by the Author for his own use in the construction of a great variety of MODERN ENGINEERING WORK.*

The information is given in a condensed and concise form, and is illustrated by upwards of 370 Woodcuts; and comprises a quantity of tabulated matter of great value to all engaged in designing, constructing, or estimating for ENGINES, BOILERS, and OTHER ENGINEERING WORK.

*** OPINIONS OF THE PRESS.

"We have kept it at hand for several weeks, referring to it as occasion arose, and we have not on a single occasion consulted its pages without finding the information of which we were in quest."—*Athenæum*.

"A thoroughly good practical handbook, which no engineer can go through without learning something that will be of service to him."—*Marine Engineer*.

"An excellent book of reference for engineers, and a valuable text-book for students of engineering."—*Scotsman*.

"This valuable manual embodies the results and experience of the leading authorities on mechanical engineering."—*Building News*

"The author has collected together a surprising quantity of rules and practical data, and has shown much judgment in the selections he has made. . . . There is no doubt that this book is one of the most useful of its kind published, and will be a very popular compendium."—*Engineer*.

"A mass of information, set down in simple language, and in such a form that it can be easily referred to at any time. The matter is uniformly good and well chosen, and is greatly elucidated by the illustrations. The book will find its way on to most engineers' shelves, where it will rank as one of the most useful books of reference."—*Practical Engineer*.

"Full of useful information, and should be found on the office shelf of all practical engineers."—*English Mechanic*.

MR. HUTTON'S PRACTICAL HANDBOOKS—*continued.*

Practical Treatise on Modern Steam-Boilers.

STEAM BOILER CONSTRUCTION. A Practical Handbook for Engineers, Boiler-Makers, and Steam Users. Containing a large Collection of Rules and Data relating to Recent Practice in the Design, Construction, and Working of all Kinds of Stationary, Locomotive, and Marine Steam-Boilers. By WALTER S. HUTTON, Civil and Mechanical Engineer, Author of "The Works' Manager's Handbook," "The Practical Engineer's Handbook," &c. With upwards of 300 Illustrations. Second Edition, medium 8vo, 18s. cloth.

☞ THIS WORK *is issued in continuation of the Series of Handbooks written by the Author, viz:—*"THE WORKS' MANAGER'S HANDBOOK" *and* "THE PRACTICAL ENGINEER'S HANDBOOK," *which are so highly appreciated by Engineers for the practical nature of their information ; and is consequently written in the same style as those works.*

The Author believes that the concentration, in a convenient form for easy reference, of such a large amount of thoroughly practical information on Steam-Boilers, will be of considerable service to those for whom it is intended, and he trusts the book may be deemed worthy of as favourable a reception as has been accorded to its predecessors.

** OPINIONS OF THE PRESS.

"Every detail, both in boiler design and management, is clearly laid before the reader. The volume shows that boiler construction has been reduced to the condition of one of the most exact sciences ; and such a book is of the utmost value to the *fin de siècle* Engineer and Works' Manager."—*Marine Engineer.*

"There has long been room for a modern handbook on steam boilers ; there is not that room now, because Mr. Hutton has filled it. It is a thoroughly practical book for those who are occupied in the construction, design, selection, or use of boilers."—*Engineer.*

"The book is of so important and comprehensive a character that it must find its way into the libraries of every one interested in boiler using or boiler manufacture if they wish to be thoroughly informed. We strongly recommend the book for the intrinsic value of its contents."—*Machinery Market.*

"The value of this book can hardly be over-estimated. The author's rules, formulæ, &c., are all very fresh, and it is impossible to turn to the work and not find what you want. No practical engineer should be without it."—*Colliery Guardian.*

Hutton's "Modernised Templeton."

THE PRACTICAL MECHANICS' WORKSHOP COMPANION. Comprising a great variety of the most useful Rules and Formulæ in Mechanical Science, with numerous Tables of Practical Data and Calculated Results for Facilitating Mechanical Operations. By WILLIAM TEMPLETON, Author of "The Engineer's Practical Assistant," &c. &c. Seventeenth Edition, Revised, Modernised, and considerably Enlarged by WALTER S. HUTTON, C.E., Author of "The Works' Manager's Handbook," "The Practical Engineer's Handbook," &c. Fcap. 8vo, nearly 500 pp., with 8 Plates and upwards of 250 Illustrative Diagrams, 6s. strongly bound for workshop or pocket wear and tear.

** OPINIONS OF THE PRESS.

"In its modernised form Hutton's 'Templeton' should have a wide sale, for it contains much valuable information which the mechanic will often find of use, and not a few tables and notes which he might look for in vain in other works. This modernised edition will be appreciated by all who have learned to value the original editions of 'Templeton.'"—*English Mechanic.*

"It has met with great success in the engineering workshop, as we can testify ; and there are a great many men who, in a great measure, owe their rise in life to this little book."—*Building News.*

"This familiar text book—well known to all mechanics and engineers—is of essential service to the every-day requirements of engineers, millwrights, and the various trades connected with engineering and building. The new modernised edition is worth its weight in gold."—*Building News.* (Second Notice.)

"This well-known and largely-used book contains information, brought up to date, of the sort so useful to the foreman and draughtsman. So much fresh information has been introduced as to constitute it practically a new book. It will be largely used in the office and workshop."—*Mechanical World.*

"The publishers wisely entrusted the task of revision of this popular, valuable, and useful book to Mr. Hutton than whom a more competent man they could not have found."—*Iron.*

Templeton's Engineer's and Machinist's Assistant.

THE ENGINEER'S, MILLWRIGHT'S, AND MACHINIST'S Practical Assistant. A collection of Useful Tables, Rules, and Data. By WILLIAM TEMPLETON. Seventh Edition, with Additions. 18mo, 2s. 6d. cloth.

"Occupies a foremost place among books of this kind. A more suitable present to an apprentice to any of the mechanical trades could not possibly be made."—*Building News.*

"A deservedly popular work. It should be in the 'drawer' of every mechanic."—*English Mechanic.*

Foley's Office Reference Book for Mechanical Engineers.

THE MECHANICAL ENGINEER'S REFERENCE BOOK, for

Machine and Boiler Construction. In Two Parts. Part I. GENERAL ENGINEERING DATA. Part II. BOILER CONSTRUCTION. With 51 Plates and numerous Illustrations. By NELSON FOLEY, M.I.N.A. Second Edition, Revised throughout and much Enlarged. Folio, £3 3s. net, half-bound.

SUMMARY OF CONTENTS.

PART I.

MEASURES.—CIRCUMFERENCES AND AREAS, &c., SQUARES, CUBES, FOURTH POWERS.—SQUARE AND CUBE ROOTS.—SURFACE OF TUBES.—RECIPROCALS.—LOGARITHMS.—MENSURATION.—SPECIFIC GRAVITIES AND WEIGHTS.—WORK AND POWER.—HEAT.—COMBUSTION.—EXPANSION AND CONTRACTION.—EXPANSION OF GASES.—STEAM.—STATIC FORCE.—GRAVITATION AND ATTRACTION.—MOTION AND COMPUTATION OF RESULTING FORCES.—ACCUMULATED WORK.—CENTRE AND RADIUS OF GYRATION.—MOMENT OF INERTIA.—CENTRE OF OSCILLATION.—ELECTRICITY.—STRENGTH OF MATERIALS.—ELASTICITY.—TEST SHEETS OF METALS.—FRICTION.—TRANSMISSION OF POWER.—FLOW OF LIQUIDS.—FLOW OF GASES.—AIR PUMPS, SURFACE CONDENSERS, &c.—SPEED OF STEAMSHIPS.—PROPELLERS.—CUTTING TOOLS.—FLANGES—COPPER SHEETS AND TUBES—SCREWS, NUTS. BOLT HEADS, &c.—VARIOUS RECIPES AND MISCELLANEOUS MATTER—WITH DIAGRAMS FOR VALVE-GEAR, BELTING AND ROPES, DISCHARGE AND SUCTION PIPES, SCREW PROPELLERS, AND COPPER PIPES.

PART II.

TREATING OF POWER OF BOILERS.—USEFUL RATIOS.—NOTES ON CONSTRUCTION.—CYLINDRICAL BOILER SHELLS.—CIRCULAR FURNACES.—FLAT PLATES.—STAYS.—GIRDERS.—SCREWS.—HYDRAULIC TESTS.—RIVETING.—BOILER SETTING, CHIMNEYS, AND MOUNTINGS.—FUELS, &c.—EXAMPLES OF BOILERS AND SPEEDS OF STEAMSHIPS.—NOMINAL AND NORMAL HORSE POWER—WITH DIAGRAMS FOR ALL BOILER CALCULATIONS AND DRAWINGS OF MANY VARIETIES OF BOILERS.

*** OPINIONS OF THE PRESS.

"The book is one which every mechanical engineer may, with advantage to himself, add to his library."—*Industries.*

"Mr. Foley is well fitted to compile such a work. . . . The diagrams are a great feature of the work. . . . Regarding the whole work, it may be very fairly stated that Mr. Foley has produced a volume which will undoubtedly fulfil the desire of the author and become indispensable to all mechanical engineers."—*Marine Engineer.*

"We have carefully examined this work, and pronounce it a most excellent reference book for the use of marine engineers."—*Journal of American Society of Naval Engineers.*

"A veritable monument of industry on the part of Mr. Foley, who has succeeded in producing what is simply invaluable to the engineering profession."—*Steamship.*

Coal and Speed Tables.

A POCKET BOOK OF COAL AND SPEED TABLES, for

Engineers and Steam-users. By NELSON FOLEY, Author of "The Mechanical Engineer's Reference Book." Pocket-size, 3s. 6d. cloth.

"These tables are designed to meet the requirements of every-day use; they are of sufficient scope for most practical purposes, and may be commended to engineers and users of steam."—*Iron.*

"This pocket-book well merits the attention of the practical engineer. Mr. Foley has compiled a very useful set of tables, the information contained in which is frequently required by engineers, coal consumers, and users of steam."—*Iron and Coal Trades Review.*

Steam Engine.

TEXT-BOOK ON THE STEAM ENGINE. With a Supplement

on GAS ENGINES, and PART II. ON HEAT ENGINES. By T. M. GOODEVE, M.A., Barrister-at-Law, Professor of Mechanics at the Royal College of Science, London; Author of "The Principles of Mechanics," "The Elements of Mechanism," &c. Thirteenth Edition. Crown 8vo, 6s. cloth.

"Professor Goodeve has given us a treatise on the steam engine, which will bear comparison with anything written by Huxley or Maxwell, and we can award it no higher praise."—*Engineer.*

"Mr. Goodeve's text-book is a work of which every young engineer should possess himself."—*Mining Journal.*

Gas Engines.

ON GAS ENGINES. With Appendix describing a Recent Engine

with Tube Igniter. By T. M. GOODEVE, M.A. Crown 8vo, 2s. 6d. cloth.

"Like all Mr. Goodeve's writings, the present is no exception in point of general excellence. It is a valuable little volume."—*Mechanical World.*

Steam Boilers.

A TREATISE ON STEAM BOILERS: Their Strength, Con-

struction, and Economical Working. By R. WILSON, C.E. Fifth Edition. 12mo, 6s. cloth.

"The best treatise that has ever been published on steam boilers."—*Engineer.*

"The author shows himself perfect master of his subject, and we heartily recommend all employing steam power to possess themselves of the work."—*Ryland's Iron Trade Circular.*

Steam Engine Design.

A HANDBOOK ON THE STEAM ENGINE, with especial Reference to Small and Medium-sized Engines. For the Use of Engine Makers, Mechanical Draughtsmen, Engineering Students, and Users of Steam Power. By HERMAN HEADER, C.E. Translated from the German with considerable Additions and Alterations, by H. H. P. POWLES, A.M.I.C.E., M.I.M.E. Second Edition, Revised. With nearly 1,100 Illustrations. Crown 8vo, 9s. cloth.

"A perfect encyclopædia of the steam engine and its details, and one which must take a permanent place in English drawing-offices and workshops."—*A Foreman Pattern-maker.*

"This is an excellent book, and should be in the hands of all who are interested in the construction and design of medium-sized stationary engines. . . . A careful study of its contents and the arrangement of the sections leads to the conclusion that there is probably no other book like it in this country. The volume aims at showing the results of practical experience, and it certainly may claim a complete achievement of this idea."—*Nature.*

"There can be no question as to its value. We cordially commend it to all concerned in the design and construction of the steam engine."—*Mechanical World.*

Boiler Chimneys.

BOILER AND FACTORY CHIMNEYS: Their Draught-Power and Stability. With a Chapter on *Lightning Conductors.* By ROBERT WILSON, A.I.C.E., Author of "A Treatise on Steam Boilers," &c. Crown 8vo, 3s. 6d. cloth.

"A valuable contribution to the literature of scientific building."—*The Builder.*

Boiler Making.

BOILER-MAKER'S READY RECKONER AND ASSISTANT. With Examples of Practical Geometry and Templating, for the Use of Platers, Smiths, and Riveters. By JOHN COURTNEY, Edited by D. K. CLARK, M.I.C.E. Third Edition, 480 pp., with 140 Illustrations. Fcap. 8vo, 7s. half-bound.

"No workman or apprentice should be without this book."—*Iron Trade Circular.*

Refrigerating Machinery.

REFRIGERATING AND ICE-MAKING MACHINERY: A Descriptive Treatise for the Use of Persons Employing Refrigerating and Ice-Making Installations, and others. By A. J. WALLIS-TAYLER, C.E., Assoc. Member Inst. C.E. With Illustrations. Crown 8vo, 7s. 6d. cloth. [*Just published.*

"Practical, explicit and profusely illustrated."—*Glasgow Herald.*

"We recommend the book, which gives the cost of various systems and illustrations showing details of parts of machinery and general arrangements of complete installations."—*Builder.*

"May be recommended as a useful description of the machinery, the processes, and of the facts, figures, and tabulated physics of refrigerating. It is one of the best compilations on the subject."—*Engineer.*

Hydraulic Machinery.

HYDRAULIC MACHINERY Employed in the Concentration and Transmission of Power. By G. CROYDON MARKS, A.M.I.C.E., A.M.I.M.E. New Edition, Enlarged. Crown 8vo. [*In the press.*

Locomotive Engine Development.

THE LOCOMOTIVE ENGINE AND ITS DEVELOPMENT. A Popular Treatise on the Gradual Improvements made in Railway Engines between 1803 and 1896. By CLEMENT E. STRETTON, C.E. Fifth Edition, Revised and Enlarged. With 120 Illustrations. Crown 8vo, 3s. 6d. cloth. [*Just published.*

"Students of railway history and all who are interested in the evolution of the modern locomotive will find much to attract and entertain in this volume."—*The Times.*

"The author of this work is well known to the railway world, and no one, probably, has a better knowledge of the history and development of the locomotive. The volume before us should be of value to all connected with the railway system of this country."—*Nature.*

Estimating for Engineering Work, &c.

ENGINEERING ESTIMATES, COSTS, AND ACCOUNTS: A Guide to Commercial Engineering. With numerous Examples of Estimates and Costs of Millwright Work, Miscellaneous Productions, Steam Engines and Steam Boilers; and a Section on the Preparation of Costs Accounts. By A GENERAL MANAGER. Second Edition. Demy 8vo, 12s. cloth. [*Just published.*

"This is an excellent and very useful book, covering subject-matter in constant requisition in every factory and workshop. . . . . The book is invaluable, not only to the young engineer, but also to the estimate department of every works."—*Builder.*

"We accord the work unqualified praise. The information is given in a plain, straightforward manner, and bears throughout evidence of the intimate practical acquaintance of the author with every phrase of commercial engineering."—*Mechanical World.*

Boiler Making.

PLATING AND BOILER MAKING : A Practical Handbook for Workshop Operations. By JOSEPH G. HORNER, A.M.I.M.E. ("Foreman Pattern Maker"), Author of "Pattern Making," &c. 380 pages, with 338 Illustrations. Crown 8vo, 7s. 6d. cloth. [*Just published.*

"The latest production from the pen of this writer is characterised by that evidence of close acquaintance with workshop methods which will render the book exceedingly acceptable to the practical hand. We have no hesitaion in commending the work as a serviceable and practical handbook on a subject which has not hitherto received much attention from those qualified to deal with it in a satisfactory manner."—*Mechanical World.*

Engineering Construction.

PATTERN-MAKING : A Practical Treatise, embracing the Main Types of Engineering Construction and including Gearing, both Hand and Machine-made, Engine Work, Sheaves and Pulleys, Pipes and Columns, Screws, Machine Parts, Pumps and Cocks, the Moulding of Patterns in Loam and Greensand, &c. ; together with the methods of Estimating the weight of Castings ; to which is added an Appendix of Tables for Workshop Reference. By JOSEPH G HORNER, A.M.I.M.E. ("Foreman Pattern Maker"). Second Edition, thoroughly Revised and much Enlarged. With 450 Illustrations. Crown 8vo, 7s. 6d. cloth.

"A well-written technical guide, evidently written by a man who understands and has practised what he has written about. . . . . We cordially recommend it to engineering students, young journeymen, and others des rous of being initiated into the mysteries of pattern-making."—*Builder.*

"More than 400 illustrations help to explain the text, which is, however, always clear and explicit, thus rendering the work an excellent *vade mecum* for the apprentice who desires to become master of his trade."—*English Mechanic.*

Dictionary of Mechanical Engineering Terms.

MECHANICAL ENGINEERING TERMS (Lockwood's Dictionary of). Embracing those current in the Drawing Office, Pattern Shop, Foundry, Fitting, Turning, Smiths', and Boiler Shops, &c. &c. Comprising upwards of 6,000 Definitions. Edited by JOSEPH G. HORNER, A.M.I.M.E. ("Foreman Pattern Maker"), Author of "Pattern Making," &c. Second Edition, Revised, with Additions. Crown 8vo, 7s. 6d. cloth.

"Just the sort of handy dictionary required by the various trades engaged in mechanical engineering. The practical engineering pupil will find the book of great value in his studies, and every foreman engineer and mechanic should have a copy."—*Building News.*

"Not merely a d ctionary, but, to a certain extent, also a most valuable guide. It strikes us as a happy idea to combine with a definition of the phrase useful information on the subject of which it treats."—*Machinery Market.*

Mill Gearing.

TOOTHED GEARING : A Practical Handbook for Offices and Work- shops. By JOSEPH HORNER, A.M.I.M.E. ("Foreman Pattern Maker"), Author of "Pattern Making," &c. With 184 Illustrations. Crown 8vo, 6s. cloth.

'We must give the book our unqualified praise for its thoroughness of treatment and we can heartily recommend it to all interested as the most practical book on the subject yet written.'—*Mechanical World.*

Fire Engineering.

FIRES, FIRE-ENGINES, AND FIRE-BRIGADES. With a History of Fire-Engines, their Construction, Use, and Management ; Remarks on Fire-Proof Buildings, and the Preservation of Life from Fire ; Statistics of the Fire Appliances in English Towns ; Foreign Fire Systems ; Hints on Fire-Brigades, &c. &c. By CHARLES F. T. YOUNG, C.E. With Illustrations, 544 pp., demy 8vo, £1 4s. cloth.

"To such of our readers as are interested in the subject of fires and fire apparatus, we can most heartily commend this book. It is really the only English work we now have upon the subject."—*Engineering.*

Motor-Cars, &c.

MOTOR CARS OR POWER CARRIAGES FOR COMMON ROADS. Containing Descriptions and Illustrations of the most notable Early and Modern Examples of Self-propelled Vehicles, by A. J. WALLIS-TAYLER, C.E., A.M.I.C.E. Crown 8vo, cloth, fully illustrated, price about 4s 6d. [*In the press,*

Stone-working Machinery.

STONE-WORKING MACHINERY, and the Rapid and Economical Conversion of Stone. With Hints on the Arrangement and Management of Stone Works. By M. Powis Bale, M.I.M.E. With Illustrations. Crown 8vo, 9s.
"The book should be in the hands of every mason or student of stonework."—*Colliery Guardian.*
"A capital handbook for all who manipulate stone for building or ornamental purposes."—*Machinery Market.*

Pump Construction and Management.

PUMPS AND PUMPING: A Handbook for Pump Users. Being Notes on Selection, Construction, and Management. By M. Powis Bale, M.I.M.E. Third Edition, Revised. Crown 8vo, 2s. 6d. cloth. [*Just published.*
"The matter is set forth as concisely as possible. In fact, condensation rather than diffuseness has been the author's aim throughout; yet he does not seem to have omitted anything likely to be of use."—*Journal of Gas Lighting.*
"Thoroughly practical and simply and clearly written."—*Glasgow Herald.*

Milling Machinery, &c.

MILLING MACHINES AND PROCESSES: A Practical Treatise on Shaping Metals by Rotary Cutters. Including Information on Making and Grinding the Cutters. By Paul N. Hasluck, Author of "Lathe-Work." With upwards of 300 Engravings. Large crown 8vo, 352 pages, 12s. 6d. cloth.
"A new departure in engineering literature. . . . We can recommend this work to all interested in milling machines; it is what it professes to be—a practical treatise."—*Engineer.*
"A capital and reliable book which will no doubt be of considerable service both to those who are already acquainted with the process as well as to those who contemplate its adoption."—*Industries.*

Turning.

LATHE-WORK: A Practical Treatise on the Tools, Appliances, and Processes employed in the Art of Turning. By Paul N. Hasluck. Fifth Edition. Crown 8vo, 5s. cloth.
"Written by a man who knows not only how work ought to be done, but who also knows how to do it, and how to convey his knowledge to others. To all turners this book would be valuable."—*Engineering.*
"We can safely recommend the work to young engineers. To the amateur it will simply be invaluable. To the student it will convey a great deal of useful information."—*Engineer.*

Screw-Cutting.

SCREW THREADS: And Methods of Producing Them. With numerous Tables and complete Directions for using Screw-Cutting Lathes. By Paul N. Hasluck, Author of "Lathe-Work," &c. With Seventy-four Illustrations. Fourth Edition, Re-written and Enlarged. Waistcoat-pocket size, 1s. 6d.
"Full of useful information, hints and practical criticism. Taps, dies, and screwing tools generally are illustrated and their action described."—*Mechanical World.*
"It is a complete compendium of all the details of the screw-cutting lathe: in fact a *multum-in-parvo* on all the subjects it treats upon."—*Carpenter and Builder.*

Smith's Tables for Mechanics, &c.

TABLES & MEMORANDA FOR MECHANICS, ENGINEERS, Architects, Builders, &c. Selected and Arranged by Francis Smith. Sixth Edition, Revised, including Electrical Tables, Formulæ, and Memoranda. Waistcoat-pocket size, 1s. 6d. limp leather. [*Just published.*
"It would, perhaps, be as difficult to make a small pocket-book selection of notes and formulæ to suit ALL engineers as it would be to make a universal medicine; but Mr. Smith's waistcoat-pocket collection may be looked upon as a successful attempt."—*Engineer.*
"The best example we have ever seen of 270 pages of useful matter packed into the dimensions of a card-case."—*Building News.* "A veritable pocket treasury of knowledge."—*Iron.*

French-English Glossary for Engineers, &c.

POCKET GLOSSARY OF TECHNICAL TERMS: English-French, French-English; with Tables suitable for the Architectural, Engineering, Manufacturing, and Nautical Professions. By John James Fletcher, Engineer and Surveyor. Second Edition, Revised and Enlarged, 200 pp. Waistcoat-pocket size, 1s. 6d. limp leather.
"It is a very great advantage for readers and correspondents in France and England to have so large a number of the words relating to engineering and manufacturers collected in a liliputian volume. The little book will be useful both to students and travellers."—*Architect.*
"The glossary of terms is very complete, and many of the Tables are new and well arranged We cordially commend the book."—*Mechanical World.*

Year-Book of Engineering Formulæ, &c.

THE ENGINEER'S YEAR-BOOK FOR 1897. Comprising Formulæ, Rules, Tables, Data and Memoranda in Civil, Mechanical, Electrical, Marine and Mine Engineering. By H. R. KEMPE, A.M.Inst.C.E., M.I.E.E., Technical Officer of the Engineer-in-Chief's Office, General Post Office, London, Author of "A Handbook of Electrical Testing," "The Electrical Engineer's Pocket-Book," &c. With about 850 Illustrations, specially Engraved for the work. Crown 8vo, 670 pages, 8s. leather. *[Just published.*

"Represents an enormous quantity of work, and forms a desirable book of reference."—*The Engineer.*
"The volume is distinctly in advance of most similar publications in this country."—*Engineering.*
"This valuable and well-designed book of reference meets the demands of all descriptions of engineers."—*Saturday Review.*
"Teems with up-to-date information in every branch of engineering and construction."—*Building News.*
"The needs of the engineering profession could hardly be supplied in a more admirable, complete and convenient form. To say that it more than sustains all comparisons is praise of the highest sort, and that may justly be said of it."—*Mining Journal.*
"There is certainly room for the new comer, which supplies explanations and directions, as well as formulæ and tables. It deserves to become one of the most successful of the technical annuals."—*Architect.*
"Brings together with great skill all the technical information which an engineer has to use day by day. It is in every way admirably equipped, and is sure to prove successful."—*Scotsman.*
"The up-to-dateness of Mr. Kempe's compilation is a quality that will not be lost on the busy people for whom the work is intended."—*Glasgow Herald.*

Portable Engines.

THE PORTABLE ENGINE: Its Construction and Management: A Practical Manual for Owners and Users of Steam Engines generally. By WILLIAM DYSON WANSBROUGH. Crown 8vo, 3s. 6d. cloth.

"This is a work of value to those who use steam machinery. . . . Should be read by every one who has a steam engine, on a farm or elsewhere."—*Mark Lane Express.*
"We cordially commend this work to buyers and owners of steam engines, and to those who have to do with their construction or use."—*Timber Trades Journal.*
"Such a general knowledge of the steam-engine as Mr. Wansbrough furnishes to the reader should be acquired by all intelligent owners and others who use the steam engine."—*Building News.*
"An excellent text-book of this useful form of engine. The 'Hints to Purchasers' contain a good deal of common-sense and practical wisdom."—*English Mechanic.*

Iron and Steel.

"IRON AND STEEL": A Work for the Forge, Foundry, Factory, and Office. Containing ready, useful, and trustworthy Information for Ironmasters and their Stock-takers; Managers of Bar, Rail, Plate, and Sheet Rolling Mills; Iron and Metal Founders; Iron Ship and Bridge Builders; Mechanical, Mining, and Consulting Engineers; Architects, Contractors, Builders, &c. By CHARLES HOARE, Author of "The Slide Rule," &c. Ninth Edition. 32mo, 6s. leather.

"For comprehensiveness the book has not its equal."—*Iron.*
"One of the best of the pocket books."—*English Mechanic.*

Elementary Mechanics.

CONDENSED MECHANICS. A Selection of Formulæ, Rules, Tables, and Data for the Use of Engineering Students, Science Classes, &c. In accordance with the Requirements of the Science and Art Department. By W. G. CRAWFORD HUGHES, A.M.I.C.E. Crown 8vo, 2s. 6d. cloth.

"The book is well fitted for those who are either confronted with practical problems in their work, or are preparing for examination and wish to refresh their knowledge by going through their formulæ again."—*Marine Engineer.*
"It is well arranged, and meets the wants of those for whom it is intended."—*Railway News.*

Steam.

THE SAFE USE OF STEAM. Containing Rules for Unprofessional Steam-users. By an ENGINEER. Seventh Edition. Sewed, 6d.

"If steam-users would but learn this little book by heart, boiler explosions would become sensations by their rarity."—*English Mechanic.*

Warming.

HEATING BY HOT WATER; with Information and Suggestions on the best Methods of Heating Public, Private and Horticultural Buildings. By WALTER JONES. Second Edition. With 96 Illustrations, crown 8vo, 2s. 6d. net.

"We confidently recommend all interested in heating by hot water to secure a copy of this valuable little treatise."—*The Plumber and Decorator.*

CIVIL ENGINEERING, SURVEYING, etc.

Light Railways.

LIGHT RAILWAYS FOR THE UNITED KINGDOM, India, and the Colonies ; A Practical Handbook setting forth the Principles on which Light Railways should be Constructed, Worked and Financed ; and detailing the cost of Construction, Equipment, Revenue and Working Expenses of Local Railways already established in the above-mentioned countries, and in Belgium, France, Switzerland &c. By J. C. MACKAY, F.G.S., A.M. Inst. C.E. Illustrated with Plates and Diagrams. Medium 8vo, 15s. cloth. *[Just published.*

"Mr. Mackay's volume is clearly and concisely written, admirably arranged, and freely illustrated. The book is exactly what has been long wanted. We recommend it to all interested in the subject. It is sure to have a wide sale."—*Railway News.*

"Those who desire to have within reach general information concerning almost all the light railway systems in the world will do well to buy Mr. Mackay's book."—*Engineer.*

"This work appears very opportunely, when the extension of the system on a large scale to England is at last being mooted. In its pages we find all the information that the heart of man can desire on the subject. . . . every detail in its story, founded on the experience of other countries and applied to the possibilities of England, is put before us."—*Spectator.*

Tunnelling.

PRACTICAL TUNNELLING. Explaining in detail Setting-out the Works, Shaft-sinking, and Heading-driving, Ranging the Lines and Levelling underground, Sub-Excavating, Timbering and the Construction of the Brickwork of Tunnels, with the amount of Labour required for, and the Cost of, the various portions of the work. By FREDERICK W. SIMMS, M.Inst.C.E. Fourth Edition, Revised and Further Extended, including the most Recent (1895) Examples of Sub-aqueous and other Tunnels by D. KINNEAR CLARK, M.Inst.C.E. Imperial 8vo, with 34 Folding Plates and other Illustrations. £2 2s., cloth. *[Just published.*

"The present (1896) edition has been brought right up to date, and is thus rendered a work to which civil engineers generally should have ready access, and to which engineers who have construction work can hardly afford to be without, but which to the younger members of the profession is invaluable, as from its pages they can learn the state to which the science of tunnelling has attained."—*Railway News.*

"The estimation in which Mr. Simms's book on tunnelling has been held for many years cannot be more truly expressed than in the words of the late Prof. Rankine : ' The best source of information on the subject of tunnels is Mr. F. W. Simms's work on Practical Tunnelling ' "—*Architect.*

Water Supply and Water-Works.

THE WATER SUPPLY OF TOWNS and the Construction of Water-Works : A Practical Treatise for the Use of Engineers and Students of Engineering. By W. K. BURTON, A.M.Inst.C.E., Professor of Sanitary Engineering in the Imperial University, Tokyo, Japan, and Consulting Engineer to the Tokyo Water-Works. With an Appendix on THE EFFECTS OF EARTHQUAKES ON WATER-WORKS, by Professor JOHN MILNE, F.R.S. With numerous Plates and Illustrations. Super-royal 8vo, 25s. buckram.

I. INTRODUCTORY. — II. DIFFERENT QUALITIES OF WATER.—III. QUANTITY OF WATER TO BE PROVIDED.—IV. ON ASCERTAINING WHETHER A PROPOSED SOURCE OF SUPPLY IS SUFFICIENT. —V. ON ESTIMATING THE STORAGE CAPACITY REQUIRED TO BE PROVIDED —VI. CLASSIFICATION OF WATERWORKS.—VII. IMPOUNDING RESERVOIRS. — VIII. EARTHWORK DAMS. — IX. MASONRY DAMS.—X. THE PURIFICATION OF WATER. — XI. SETTLING RESERVOIRS — XII. SAND FILTRATION.— XIII. PURIFICATION OF WATER BY ACTION OF IRON, SOFTENING OF WATER BY ACTION OF LIME, NATURAL FILTRATION — XIV.—SERVICE OR CLEAN WATER RESERVOIRS— WATER TOWERS—STAND PIPES.—XV. THE CONNECTION OF SETTLING RESERVOIRS, FILTER BEDS AND SERVICE RESERVOIRS.—XVI. PUMPING MACHINERY.—XVII. FLOW OF WATER IN CONDUITS —PIPES AND OPEN CHANNELS.—XVIII. DISTRIBUTION SYSTEMS.- XIX. SPECIAL PROVISIONS FOR THE EXTINCTION OF FIRE.—XX. PIPES FOR WATERWORKS.—XXI. PREVENTION OF WASTE OF WATER. — XXII. VARIOUS APPLICATIONS USED IN CONNECTION WITH WATERWORKS.

APPENDIX. By PROF. JOHN MILNE, F.R.S. —CONSIDERATIONS CONCERNING THE PROBABLE EFFECTS OF EARTHQUAKES ON WATERWORKS, AND THE SPECIAL PRECAUTIONS TO BE TAKEN IN EARTHQUAKE COUNTRIES.

"The chapter upon filtration of water is very complete, and the details of construction well illustrated. . . The work should be specially valuable to civil engineers engaged in work in Japan, but the interest is by no means confined to that locality."—*Engineer.*

"We congratulate the author upon the practical commonsense shown in the preparation of this work. . . . The plates and diagrams have evidently been prepared with great care, and cannot fail to be of great assistance to the student."—*Builder.*

"The whole art of waterworks construction is dealt with in a clear and comprehensive fashion in this handsome volume. . . . Mr. Burton's practical treatise shows in all its sections the fruit of independent study and individual experience. It is largely based upon his own practice in the branch of engineering of which it treats."—*Saturday Review*

The Water-Supply of Cities and Towns.

THE WATER SUPPLY OF CITIES AND TOWNS (A Comprehensive Treatise on, by WILLIAM HUMBER, A.-M.Inst.C.E., and M.Inst.M.E., Author of "A Complete Treatise on Cast and Wrought Iron Bridge Construction," &c. &c. Illustrated with 50 Double Plates, 1 Single Plate, Coloured Frontispiece, and upwards of 250 Woodcuts, and containing 400 pages of Text. Imp. 4to, £6 6s. elegantly and substantially half-bound in morocco.

LIST OF CONTENTS.

I. HISTORICAL SKETCH OF SOME OF THE MEANS THAT HAVE BEEN ADOPTED FOR THE SUPPLY OF WATER TO CITIES AND TOWNS.—II. WATER AND THE FOREIGN MATTER USUALLY ASSOCIATED WITH IT.—III. RAINFALL AND EVAPORATION.—IV. SPRINGS AND THE WATER BEARING FORMATIONS OF VARIOUS DISTRICTS.—V. MEASUREMENT AND ESTIMATION OF THE FLOW OF WATER.—VI. ON THE SELECTION OF THE SOURCE OF SUPPLY.—VII. WELLS—VIII. RESERVOIRS.—IX. THE PURIFICATION OF WATER.—X. PUMPS.—XI. PUMPING MACHINERY.—XII. CONDUITS.— | XIII. DISTRIBUTION OF WATER.—XIV. METERS, SERVICE PIPES, AND HOUSE FITTINGS.—XV. THE LAW AND ECONOMY OF WATER WORKS.—XVI. CONSTANT AND INTERMITTENT SUPPLY.—XVII. DESCRIPTION OF PLATES.—APPENDICES, GIVING TABLES OF RATES OF SUPPLY, VELOCITIES, &C. &C., TOGETHER WITH SPECIFICATIONS OF SEVERAL WORKS ILLUSTRATED, AMONG WHICH WILL BE FOUND: ABERDEEN, BIDEFORD, CANTERBURY, DUNDEE, HALIFAX, LAMBETH ROTHERHAM, DUBLIN, AND OTHERS.

"The most systematic and valuable work upon water supply hitherto produced in English, or in any other language. . . . Mr. Humber's work is characterised almost throughout by an exhaustiveness much more distinctive of French and German than of English technical treatises."—*Engineer.*

Water Supply.

RURAL WATER SUPPLY: A Practical Handbook on the Supply of Water and Construction of Waterworks for small Country Districts. By ALLAN GREENWELL, A.M.I.C.E., and W. T. CURRY, A.M.I.C.E., F.G.S. With Illustrations. Crown 8vo, 5s. cloth. [*Just published.*

"We conscientiously recommend it as a very useful book for those concerned in obtaining water for small districts, giving a great deal of practical information in a small compass."—*Builder.*
"The volume contains valuable information upon all matters connected with water supply. . . It is full of details on points which are continually before waterworks engineers."—*Nature.*

Hydraulic Tables.

HYDRAULIC TABLES, CO-EFFICIENTS, AND FORMULÆ for Finding the Discharge of Water from Orifices, Notches, Weirs, Pipes, and Rivers. With New Formulæ, Tables, and General Information on Rain-fall, Catchment-Basins, Drainage, Sewerage, Water Supply for Towns and Mill Power. By JOHN NEVILLE, Civil Engineer, M.R.I.A. Third Edition, carefully revised, with considerable Additions. Numerous Illustrations. Crown 8vo, 14s. cloth.

"It is, of all English books on the subject, the one nearest to completeness . . . . From the good arrangement of the matter, the clear explanations and abundance of formulæ, the carefully calculated tables, and, above all, the thorough acquaintance with both theory and construction, which is displayed from first to last, the book will be found to be an acquisition."—*Architect.*

Hydraulics.

HYDRAULIC MANUAL. Consisting of Working Tables and Explanatory Text. Intended as a Guide in Hydraulic Calculations and Field Operations. By LOWIS D'A. JACKSON, Author of "Aid to Survey Practice," "Modern Metrology," &c. Fourth Edition, Enlarged. Large crown 8vo, 16s. cloth.

"The author has had a wide experience in hydraulic engineering and has been a careful observer of the facts which have come under his notice, and from the great mass of material at his command he has constructed a manual which may be accepted as a trustworthy guide to this branch of the engineer's profession."—*Engineering.*
"The most useful feature of this work is its freedom from what is superannuated, and its thorough adoption of recent experiments; the text is in fact in great part a short account of the great modern experiments."—*Nature.*

Water Storage, Conveyance, and Utilisation.

WATER ENGINEERING: A Practical Treatise on the Measurement, Storage, Conveyance, and Utilisation of Water for the Supply of Towns, for Mill Power, and for other Purposes. By CHARLES SLAGG, A.-M.Inst.C.E. Second Edition. Crown 8vo, 7s. 6d. cloth.

"As a small practical treatise on the water supply of towns, and on some applications of water-power, the work is in many respects excellent."—*Engineering.*
"The author has collated the results deduced from the experiments of the most eminent authorities, and has presented them in a compact and practical form, accompanied by very clear and detailed explanations. . . The application of water as a motive power is treated very carefully and exhaustively."—*Builder.*

Drainage.

DRAINAGE OF LANDS, TOWNS, AND BUILDINGS. By
G. D. DEMPSEY, C.E. Revised, with large Additions on RECENT PRACTICE IN DRAINAGE ENGINEERING, by D. KINNEAR CLARK, M.Inst. C.E., Author of "Tramways : their Construction and Working," "A Manual of Rules, Tables, and Data for Mechanical Engineers," &c. Third Edition. Fcap. 8vo, 4s. 6d. cloth.

"The new matter added to Mr. Dempsey's excellent work is characterised by the comprehensive grasp and accuracy of detail for which the name of Mr. D. K. Clark is a sufficient voucher."—*Athenæum.*

"As a work on recent practice in drainage engineering, the book is to be commended to all who are making that branch of engineering science their special study."—*Iron.*

"A comprehensive manual on drainage engineering, and a useful introduction to the student."—*Building News.*

River Engineering.

RIVER BARS : The Causes of their Formation, and their Treatment
by "Induced Tidal Scour;" with a Description of the Successful Reduction by this Method of the Bar at Dublin. By I. J. MANN, Assist. Eng. to the Dublin Port and Docks Board. Royal 8vo, 7s. 6d. cloth.

"We recommend all interested in harbour works—and, indeed, those concerned in the improvements of rivers generally—to read Mr. Mann's interesting work on the treatment of river bars."—*Engineer.*

Tramways and their Working.

TRAMWAYS: THEIR CONSTRUCTION AND WORKING.
Embracing a Comprehensive History of the System ; with an exhaustive Analysis of the Various Modes of Traction, including Horse Power, Steam, Cable Traction, Electric Traction, &c. ; a Description of the Varieties of Rolling Stock ; and ample Details of Cost and Working Expenses. New Edition, Thoroughly Revised, and Including the Progress recently made in Tramway Construction, &c. &c. By D. KINNEAR CLARK, M.Inst. C.E. With 400 Illustrations. 8vo, 780 pages. Price 28s., buckram. [*Just published.*

"Although described as a new edition, this book is really a new one, a large part of it, which covers historical ground, having been re-written and amplified ; while the parts which relate to all that has been done since 1882 appear in this edition only. It is sixteen years since the first edition appeared, and twelve years since the supplementary volume to the first book was published. After a lapse, then, of twelve years, it is obvious that the author has at his disposal a vast quantity of descriptive and statistical information, with which he may, and has, produced a volume of great value to all interested in tramway construction and working. The new volume is one which will rank, among tramway engineers and those interested in tramway working, with his world-famed book on railway machinery."—*The Engineer*, March 8, 1895.

Student's Text-Book on Surveying.

PRACTICAL SURVEYING : A Text-Book for Students preparing
for Examinations or for Survey-work in the Colonies. By GEORGE W. USILL, A.M.I.C.E., Author of "The Statistics of the Water Supply of Great Britain." With 4 Lithographic Plates and upwards of 330 Illustrations. Fourth Edition, Revised and Enlarged. Including Tables of Natural Sines, Tangents, Secants, &c. Crown 8vo, 7s. 6d. cloth ; or, on THIN PAPER, bound in limp leather, gilt edges, rounded corners, for pocket use, price 12s. 6d.

"The best forms of instruments are described as to their construction, uses and modes of employment, and there are innumerable hints on work and equipment such as the author, in his experience as surveyor, draughtsman and teacher, has found necessary, and which the student in his inexperience will find most serviceable."—*Engineer.*

"The latest treatise in the English language on surveying, and we have no hesitation in saying that the student will find it a better guide than any of its predecessors. . . . Deserves to be recognised as the first book which should be put in the hands of a pupil of Civil Engineering, and every gentleman of education who sets out for the Colonies would find it well to have a copy."—*Architect.*

Survey Practice.

AID TO SURVEY PRACTICE : for Reference in Surveying,
Levelling, and Setting-out ; and in Route Surveys of Travellers by Land and Sea. With Tables, Illustrations, and Records. By LOWIS D'A. JACKSON, A.M.I.C.E. Second Edition, Enlarged. Large crown 8vo, 12s. 6d. cloth.

"Mr. Jackson has produced a valuable *vade-mecum* for the surveyor. We can recommend this book as containing an admirable supplement to the teaching of the accomplished surveyor."—*Athenæum.*

"As a text-book we should advise all surveyors to place it in their libraries, and study well the matured instructions afforded in its pages."—*Colliery Guardian.*

"The author brings to his work a fortunate union of theory and practical experience which, aided by a clear and lucid style of writing, renders the book a very useful one."—*Builder.*

Engineers' Field-Book.

FIELD-BOOK FOR ENGINEERS, MINING SURVEYORS, &c.
Consisting of a Series of Tables, with Rules, Explanations of Systems, and use of Theodolite for Traverse Surveying and Plotting the Work with minute accuracy by means of Straight Edge and Set Square only; Levelling with the Theodolite, Casting-out and Reducing Levels to Datum, and Plotting Sections in the ordinary manner; Setting-out Curves with the Theodolite by Tangential Angles and Multiples with Right and Left-hand Readings of the Instrument; Setting-out Curves without Theodolite on the System of Tangential Angles by Sets of Tangents and Offsets; and Earthwork Tables to 80 feet deep, calculated for every 6 inches in depth. By W. DAVIS HASKOLL, C.E. With numerous Woodcuts. Fourth Edition, Enlarged. Crown 8vo, 12s. cloth.

"The book is very handy; the separate tables of sines and tangents to every minute will make it useful for many other purposes, the genuine traverse tables existing all the same."—*Athenæum.*

"Every person engaged in engineering field operations will estimate the importance of such a work and the amount of valuable time which will be saved by reference to a set of reliable tables prepared with the accuracy and fulness of those given in this volume."—*Railway News.*

Surveying, Land and Marine.

LAND AND MARINE SURVEYING, in Reference to the Preparation of Plans for Roads and Railways; Canals, Rivers, Towns' Water Supplies; Docks and Harbours. With Description and Use of Surveying Instruments. By W. DAVIS HASKOLL, C.E., Author of "Bridge and Viaduct Construction," &c. Second Edition, Revised, with Additions. Large crown 8vo, 9s. cloth.

"This book must prove of great value to the student. We have no hesitation in recommending it, feeling assured that it will more than repay a careful study."—*Mechanical World.*

"A most useful and well arranged oook for the aid of a student. We can strongly recommend it as a carefully-written and valuable text-book. It enjoys a well-deserved repute among surveyors."—*Builder.*

"This volume cannot fail to prove of the utmost practical utility. It may be safely recommended to all students who aspire to become clean and expert surveyors."—*Mining Journal.*

Levelling.

PRINCIPLES AND PRACTICE OF LEVELLING. Showing its Application to purposes of Railway and Civil Engineering in the Construction of Roads; with Mr. TELFORD'S Rules for the same. By FREDERICK W. SIMMS, F.G.S., M. Inst. C.E. Seventh Edition, with the addition by LAW'S Practical Examples for Setting-out Railway Curves, and TRAUTWINE'S Field Practice of Laying-out Circular Curves. With 7 Plates and numerous Woodcuts, 8vo, 8s. 6d. cloth.   *** TRAUTWINE on CURVES may be had separate, 5s.

"The text-book on levelling in most of our engineering schools and colleges."—*Engineer.*

"The publishers have rendered a substantial service to the profession, especially to the younger members, by bringing out the present edition of Mr. Simms's useful work."—*Engineering.*

Trigonometrical Surveying.

AN OUTLINE OF THE METHOD OF CONDUCTING A TRIGONOMETRICAL SURVEY, for the Formation of Geographical and Topographical Maps and Plans, Military Reconnaissance, LEVELLING, &c., with Useful Problems, Formulæ, and Tables. By Lieut.-General FROME, R.E. Fourth Edition, Revised and partly Re-written by Major-General Sir CHARLES WARREN, G.C.M.G., R.E. With 19 Plates and 115 Woodcuts, royal 8vo, 16s. cloth.

"No words of praise from us can strengthen the position so well and so steadily maintained by this work. Sir Charles Warren has revised the entire work, and made such additions as were necessary to bring every portion of the contents up to the present date."—*Broad Arrow.*

Curves, Tables for Setting-out.

TABLES OF TANGENTIAL ANGLES AND MULTIPLES FOR SETTING-OUT CURVES from 5 to 200 Radius. By A. BEAZELEY, M.Inst.C.E. 4th Edition. Printed on 48 Cards, and sold in a cloth box, waistcoat-pocket size, 3s. 6d.

"Each table is printed on a small card, which, being placed on the theodolite, leaves the hands fre to manipulate the instrument—no small advantage as regards the rapidity of work."—*Engineer.*

"Very handy: a man may know that all his day's work must fall on two of these cards, which he pu into his own card-case, and leaves the rest behind."—*Athenæum.*

Earthwork.

HANDY GENERAL EARTHWORK TABLES. Giving th Contents in Cubic Yards of Centre and Slopes of Cuttings and Embankments fro 3 inches to 80 feet in Depth or Height, for use with either 66 feet Chain or 100 fee Chain. By J. H. WATSON BUCK, M.Inst.C.E. On a Sheet mounted in clot case, 3s. 6d.   *[Just published*

Earthwork.

EARTHWORK TABLES. Showing the Contents in Cubic Yards of Embankments, Cuttings, &c., of Heights or Depths up to an average of 80 feet. By JOSEPH BROADBENT, C.E., and FRANCIS CAMPIN, C.E. Crown 8vo, 5s. cloth.

" The way in which accuracy is attained, by a simple division of each cross section into three elements, two in which are constant and one variable, is ingenious."—*Athenæum.*

Earthwork, Measurement of.

A MANUAL ON EARTHWORK. By ALEX. J. S. GRAHAM, C.E. With numerous Diagrams. Second Edition. 18mo, 2s. 6d. cloth.

Tunnel Shafts.

THE CONSTRUCTION OF LARGE TUNNEL SHAFTS: A Practical and Theoretical Essay. By J. H. WATSON BUCK, M. Inst. C.E., Resident Engineer, L. and N. W. R. With Folding Plates, 8vo, 12s. cloth.

" Many of the methods given are of extreme practical value to the mason, and the observations on the form of arch, the rules for ordering the stone, and the construction of the templates, will be found of considerable use. We commend the book to the engineering profession."—*Building News.*

" Will be regarded by civil engineers as of the utmost value, and calculated to save much time and obviate many mistakes."—*Colliery Guardian.*

Cast and Wrought Iron Bridge Construction.

CAST AND WROUGHT IRON BRIDGE CONSTRUCTION (A Complete and Practical Treatise on), including Iron Foundations. In Three Parts —Theoretical, Practical, and Descriptive. By WILLIAM HUMBER, A.-M. Inst. C.E., and M. Inst. M.E. Third Edition, revised and much improved, with 115 Double Plates (20 of which now first appear in this edition), and numerous Additions to the Text. In 2 vols., imp. 4to, £6 16s. 6d. half-bound in morocco.

" A very valuable contribution to the standard literature of civil engineering. In addition to elevations, plans, and sections, large scale details are given, which very much enhance the instructive worth of those illustrations."—*Civil Engineer and Architect's Journal.*

" Mr. Humber's stately volumes, lately issued—in which the most important bridges erected during the last five years, under the direction of the late Mr. Brunel, Sir W. Cubitt, Mr. Hawkshaw, Mr. Page, Mr. Fowler, Mr. Hemans, and others among our most eminent engineers, are drawn and specified in great detail."—*Engineer.*

Oblique Bridges.

ESSAY ON OBLIQUE BRIDGES (Practical and Theoretical). With 13 large Plates. By the late GEORGE WATSON BUCK, M.I.C.E. Fourth Edition, revised by his Son, J. H. WATSON BUCK, M.I.C.E.; and with the addition of Description to Diagrams for Facilitating the Construction of Oblique Bridges, by W. H. BARLOW, M.I.C.E. Royal 8vo, 12s. cloth.

" The standard text-book for all engineers regarding skew arches is Mr. Buck's treatise, and it would be impossible to consult a better."—*Engineer.*

" Mr. Buck's treatise is recognised as a standard text-book, and his treatment has divested the subject of many of the intricacies supposed to belong to it. As a guide to the engineer and architect, on a confessedly difficult subject, Mr. Buck's work is unsurpassed."—*Building News.*

Oblique Arches.

THE CONSTRUCTION OF OBLIQUE ARCHES (A Practical Treatise on). By JOHN HART. Third Edition, with Plates. Imperial 8vo, 8s. cloth.

Statics, Graphic and Analytic.

GRAPHIC AND ANALYTIC STATICS, in their Practical Application to the Treatment of Stresses in Roofs, Solid Girders, Lattice, Bowstring, and Suspension Bridges, Braced Iron Arches and Piers, and other Frameworks. By R. HUDSON GRAHAM, C.E. Containing Diagrams and Plates to Scale. With numerous Examples, many taken from existing Structures. Specially arranged for Class-work in Colleges and Universities. Second Edition, Revised and Enlarged. 8vo, 16s. cloth.

" Mr. Graham's book will find a place wherever graphic and analytic statics are used or studied."— *Engineer.*

" The work is excellent from a practical point of view, and has evidently been prepared with much care. The directions for working are ample, and are illustrated by an abundance of well-selected examples. It is an excellent text-book for the practical draughtsman."—*Athenæum.*

Girders, Strength of.

GRAPHIC TABLE for Facilitating the Computation of the Weights of Wrought Iron and Steel Girders, &c., for Parliamentary and other Estimates. By J. H. WATSON BUCK, M. Inst. C.E. On a Sheet, 2s. 6d.

Strains, Calculation of.

A HANDY BOOK FOR THE CALCULATION OF STRAINS in Girders and Similar Structures and their Strength. Consisting of Formulæ and Corresponding Diagrams, with numerous details for Practical Application, &c. By WILLIAM HUMBER, A.-M.Inst.C.E., &c. Fifth Edition. Crown 8vo, with nearly 100 Woodcuts and 3 Plates, 7s. 6d. cloth.

"The formulæ are neatly expressed, and the diagrams good."—*Athenæum.*
"We heartily commend this really *handy* book to our engineer and architect readers."—*English Mechanic.*

Trusses.

TRUSSES OF WOOD AND IRON. Practical Applications of Science in Determining the Stresses, Breaking Weights, Safe Loads, Scantlings, and Details of Construction. With Complete Working Drawings. By WILLIAM GRIFFITHS, Surveyor, Assistant Master, Tranmere School of Science and Art. Oblong 8vo, 4s. 6d. cloth.

"This handy little book enters so minutely into every detail connected with the construction of roof trusses that no student need be ignorant of these matters."—*Practical Engineer.*

Strains in Ironwork.

THE STRAINS ON STRUCTURES OF IRONWORK; with Practical Remarks on Iron Construction. By F. W. SHEILDS, M.I.C.E. 8vo, 5s. cl.

Barlow's Strength of Materials, Enlarged by Humber.

A TREATISE ON THE STRENGTH OF MATERIALS; with Rules for application in Architecture, the Construction of Suspension Bridges, Railways, &c. By PETER BARLOW, F.R.S. A New Edition, revised by his Sons, P. W. BARLOW, F.R.S., and W. H. BARLOW, F.R.S.; to which are added, Experiments by HODGKINSON, FAIRBAIRN, and KIRKALDY; and Formulæ for Calculating Girders, &c. Arranged and Edited by WM. HUMBER, A.-M.Inst.C.E. Demy 8vo, 400 pp., with 19 large Plates and numerous Woodcuts, 18s. cloth.

"Valuable alike to the student, tyro, and the experienced practitioner, it will always rank in future as it has hitherto done, as the standard treatise on that particular subject."—*Engineer.*
"As a scientific work of the first class, it deserves a foremost place on the bookshelves of every civil engineer and practical mechanic."—*English Mechanic.*

Cast Iron and other Metals, Strength of.

STRENGTH OF CAST IRON AND OTHER METALS. By THOMAS TREDGOLD, C.E. Fifth Edition, including HODGKINSON'S Experimental Researches. 8vo, 12s. cloth.

Railway Working.

SAFE RAILWAY WORKING: A Treatise on Railway Accidents, their Cause and Prevention; with a Description of Modern Appliances and Systems. By CLEMENT E. STRETTON, C.E., Vice-President and Consulting Engineer, Amalgamated Society of Railway Servants. With Illustrations and Coloured Plates. Third Edition, Enlarged. Crown 8vo, 3s. 6d. cloth.

"A book for the engineer, the directors, the managers; and, in short, all who wish for information on railway matters will find a perfect encyclopædia in 'Safe Railway Working.'"—*Railway Review.*
"We commend the remarks on railway signalling to all railway managers, especially where a uniform code and practice is advocated."—*Herepath's Railway Journal.*
"The author may be congratulated on having collected, in a very convenient form, much valuable information on the principal questions affecting the safe working of railways."—*Railway Engineer.*

Heat, Expansion by:

EXPANSION OF STRUCTURES BY HEAT. By JOHN KEILY, C.E., late of the Indian Public Works Department. Crown 8vo, 3s. 6d. cloth.

"The aim the author has set before him, viz., to show the effects of heat upon metallic and other structures, is a laudable one, for this is a branch of physics upon which the engineer or architect can find but little reliable and comprehensive data in books."—*Builder.*

Field Fortification.

A TREATISE ON FIELD FORTIFICATION, The Attack o Fortresses, Military Mining, and Reconnoitring. By Professor Colonel I. S. MACAULAY. Sixth Edition, crown 8vo, with separate Atlas of 12 Plates, 12s. cloth

MR. HUMBER'S GREAT WORK ON MODERN ENGINEERING.

Complete in Four Volumes, imperial 4to, price £12 12s. half-morocco. Each volume sold separately as follows :—

RECORD OF THE PROGRESS OF MODERN ENGINEERING.

FIRST SERIES. Comprising Civil, Mechanical, Marine, Hydraulic, Railway, Bridge, and other Engineering Works, &c. By WILLIAM HUMBER, A. M. Inst C.E., &c. Imp. 4to, with 36 Double Plates, drawn to a large scale, Photographic Portrait of John Hawkshaw, C.E., F.R.S., &c., and copious descriptive Letterpress, Specifications, &c., £3 3s. half-morocco.

LIST OF THE PLATES AND DIAGRAMS.

VICTORIA STATION AND ROOF, L. B. & S. C. R. (8 PLATES); SOUTHPORT PIER (2 PLATES); VICTORIA STATION AND ROOF, L. C. & D. AND G. W. R. (6 PLATES); ROOF OF CREMORNE MUSIC HALL; BRIDGE OVER G. N. RAILWAY; ROOF OF STATION, DUTCH RHENISH RAIL (2 PLATES); BRIDGE OVER THE THAMES, WEST LONDON EXTENSION RAILWAY (5 PLATES); ARMOUR PLATES: SUSPENSION BRIDGE, THAMES (4 PLATES); THE ALLEN ENGINE; SUSPENSION BRIDGE, AVON (3 PLATES); UNDERGROUND RAILWAY (3 PLATES).

"Handsomely lithographed and printed. It will find favour with many who desire to preserve in a permanent form copies of the plans and specifications prepared for the guidance of the contractors for many important engineering works."—*Engineer.*

HUMBER'S PROGRESS OF MODERN ENGINEERING. SECOND

SERIES. Imp. 4to, with 3 Double Plates, Photographic Portrait of Robert Stephenson, C.E., M.P., F.R.S., &c., and copious descriptive Letterpress, Specifications, &c., £3 3s. half-morocco.

LIST OF THE PLATES AND DIAGRAMS.

BIRKENHEAD DOCKS, LOW WATER BASIN (15 PLATES); CHARING CROSS STATION ROOF, C. C. RAILWAY (3 PLATES); DIGSWELL VIADUCT, GREAT NORTHERN RAILWAY; ROBBERY WOOD VIADUCT, GREAT NORTHERN RAILWAY; IRON PERMANENT WAY; CLYDACH VIADUCT; MERTHYR, TREDEGAR, AND ABERGAVENNY RAILWAY; EBBW VIADUCT, MERTHYR, TREDEGAR, AND ABERGAVENNY RAILWAY; COLLEGE WOOD VIADUCT, CORNWALL RAILWAY; DUBLIN WINTER PALACE ROOF (3 PLATES); BRIDGE OVER THE THAMES, L. C. and D. RAILWAY (6 PLATES); ALBERT HARBOUR, GREENOCK (4 PLATES).

"Mr. Humber has done the profession good and true service, by the fine collection of examples he has here brought before the profession and the public."—*Practical Mechanic's Journal.*

HUMBER'S PROGRESS OF MODERN ENGINEERING. THIRD

SERIES. Imp. 4to, with 40 Double Plates, Photographic Portrait of J. R. M'Clean, late Pres. Inst. C.E., and copious descriptive Letterpress, Specifications, &c., £3 3s. half-morocco.

LIST OF THE PLATES AND DIAGRAMS.

MAIN DRAINAGE, METROPOLIS.—*North Side.*—MAP SHOWING INTERCEPTION OF SEWERS; MIDDLE LEVEL SEWER (2 PLATES); OUTFALL SEWER, BRIDGE OVER RIVER LEA (3 PLATES); OUTFALL SEWER, BRIDGE OVER MARSH LANE, NORTH WOOLWICH RAILWAY, AND BOW AND BARKING RAILWAY JUNCTION; OUTFALL SEWER, BRIDGE OVER BOW AND BARKING RAILWAY (3 PLATES); OUTFALL SEWER, BRIDGE OVER EAST LONDON WATERWORKS' FEEDER (2 PLATES); OUTFALL SEWER RESERVOIR (2 PLATES); OUTFALL SEWER, TUMBLING BAY AND OUTLET; OUTFALL SEWER, PENSTOCKS. *South Side.*—OUTFALL SEWER, BERMONDSEY BRANCH (2 PLATES); OUTFALL SEWER, RESERVOIR AND OUTLET (4 PLATES); OUTFALL SEWER, FILTH HOIST; SECTIONS OF SEWERS (NORTH AND SOUTH SIDES). THAMES EMBANKMENT.—SECTION OF RIVER WALL; STEAMBOAT PIER, WESTMINSTER (2 PLATES); LANDING STAIRS BETWEEN CHARING CROSS AND WATERLOO BRIDGES; YORK GATE (2 PLATES); OVERFLOW AND OUTLET AT SAVOY STREET SEWER (3 PLATES); STEAMBOAT PIER, WATERLOO BRIDGE (3 PLATES); JUNCTION OF SEWERS, PLANS AND SECTIONS; GULLIES, PLANS, AND SECTIONS; ROLLING STOCK; GRANITE AND IRON FORTS.

"The drawings have a constantly increasing value, and whoever desires to possess clear representations of the two great works carried out by our Metropolitan Board will obtain Mr. Humber's volume."—*Engineer.*

HUMBER'S PROGRESS OF MODERN ENGINEERING. FOURTH

SERIES. Imp. 4to, with 36 Double Plates, Photographic Portrait of John Fowler, late Pres. Inst. C.E., and copious descriptive Letterpress, Specifications, &c., £3 3s. half-morocco.

LIST OF THE PLATES AND DIAGRAMS.

ABBEY MILLS PUMPING STATION, MAIN DRAINAGE, METROPOLIS (4 PLATES); BARROW DOCKS (5 PLATES); MANQUIS VIADUCT, SANTIAGO AND VALPARAISO RAILWAY (2 PLATES); ADAM'S LOCOMOTIVE, ST. HELEN'S CANAL RAILWAY (2 PLATES); CANNON STREET STATION ROOF, CHARING CROSS RAILWAY (3 PLATES); ROAD BRIDGE OVER THE RIVER MOKA (2 PLATES); TELEGRAPHIC APPARATUS FOR MESOPOTAMIA; VIADUCT OVER THE RIVER WYE, MIDLAND RAILWAY (3 PLATES); ST. GERMANS VIADUCT, CORNWALL RAILWAY (1 PLATES); WROUGHT-IRON CYLINDER FOR DIVING BELL; MILLWALL DOCKS (6 PLATES); MILROY'S PATENT EXCAVATOR; METROPOLITAN DISTRICT RAILWAY (6 PLATES); HARBOURS, PORTS, AND BREAKWATERS (3 PLATES).

"We gladly welcome another year's issue of this valuable publication from the able pen of Mr. Humber. The accuracy and general excellence of this work are well known, while its usefulness in giving the measurements and details of some of the latest examples of engineering, as carried out by the most eminent men in the profession, cannot be too highly prized."—*Artisan.*

THE POPULAR WORKS OF MICHAEL REYNOLDS
("THE ENGINE DRIVER'S FRIEND").

Locomotive-Engine Driving.

LOCOMOTIVE ENGINE DRIVING: A Practical Manual for Engineers in Charge of 'Locomotive Engines. By MICHAEL REYNOLDS, Member of the Society of Engineers, formerly Locomotive Inspector, L. B. and S. C. R. Ninth Edition. Including a KEY TO THE LOCOMOTIVE ENGINE. With Illustrations and Portrait of Author. Crown 8vo, 4s. 6d. cloth.

"Mr. Reynolds has supplied a want, and has supplied it well. We can confidently recommend the book not only to the practical driver, but to everyone who takes an interest in the performance of locomotive engines."—*The Engineer.*
"Mr. Reynolds has opened a new chapter in the literature of the day. This admirable practical treatise, of the practical utility of which we have to speak in terms of warm commendation."—*Athenæum.*
"Evidently the work of one who knows his subject thoroughly."—*Railway Service Gazette.*
"Were the cautions and rules given in the book to become part of the every-day working of our engine-drivers, we might have fewer distressing accidents to deplore."—*Scotsman.*

Stationary Engine Driving.

STATIONARY ENGINE DRIVING: A Practical Manual for Engineers in Charge of Stationary Engines. By MICHAEL REYNOLDS. Fifth Edition, Enlarged. With Plates and Woodcuts. Crown 8vo, 4s. 6d. cloth.

"The author is thoroughly acquainted with his subjects, and his advice on the various points treated is clear and practical. . . . He has produced a manual which is an exceedingly useful one for the class for whom it is specially intended."—*Engineering.*
"Our author leaves no stone unturned. He is determined that his readers shall not only know something about the stationary engine, but all about it."—*Engineer.*
"An engineman who has mastered the contents of Mr. Reynolds's book will require but little actual experience with boilers and engines before he can be trusted to look after them."—*English Mechanic.*

The Engineer, Fireman, and Engine-Boy.

THE MODEL LOCOMOTIVE ENGINEER, Fireman, and Engine-Boy. Comprising a Historical Notice of the Pioneer Locomotive Engines and their Inventors. By MICHAEL REYNOLDS. Second Edition, with Revised Appendix. With numerous Illustrations, and Portrait of George Stephenson. Crown 8vo, 4s. 6d. cloth. [*Just published.*

"From the technical knowledge of the author, it will appeal to the railway man of to-day more forcibly than anything written by Dr. Smiles. . . . . The volume contains information of a technical kind, and facts that every driver should be familiar with."—*English Mechanic.*
"We should be glad to see this book in the possession of everyone in the kingdom who has ever laid, or is to lay, hands on a locomotive engine."—*Iron.*

Continuous Railway Brakes.

CONTINUOUS RAILWAY BRAKES: A Practical Treatise on the several Systems in Use in the United Kingdom: their Construction and Performance. With copious Illustrations and numerous Tables. By MICHAEL REYNOLDS. Large crown 8vo, 9s. cloth.

"A popular explanation of the different brakes. It will be of great assistance in forming public opinion, and will be studied with benefit by those who take an interest in the brake."—*English Mechanic.*
"Written with sufficient technical detail to enable the principal and relative connection of the various parts of each particular brake to be readily grasped."—*Mechanical World.*

Engine-Driving Life.

ENGINE-DRIVING LIFE: Stirring Adventures and Incidents in the Lives of Locomotive Engine-Drivers. By MICHAEL REYNOLDS. Third and Cheaper Edition. Crown 8vo, 1s. 6d. cloth.

"From first to last perfectly fascinating. Wilkie Collins's most thrilling conceptions are thrown into the shade by true incidents, endless in their variety, related in every page."—*North British Mail.*
"Anyone who wishes to get a real insight into railway life cannot do better than read 'Engine-Driving Life' for himself, and if he once takes it up he will find that the author's enthusiasm and real love of the engine-driving profession will carry him on till he has read every page."—*Saturday Review.*

Pocket Companion for Enginemen.

THE ENGINEMAN'S POCKET COMPANION and Practical Educator for Enginemen, Boiler Attendants, and Mechanics. By MICHAEL REYNOLDS. With Forty-five Illustrations and numerous Diagrams. Third Edition, Revised. Royal 18mo, 3s. 6d. strongly bound for pocket wear.

"This admirable work is well suited to accomplish its object, being the honest workmanship of a competent engineer."—*Glasgow Herald.*
"A most meritorious work, giving in a succinct and practical form all the information an engine-minder desirous of mastering the scientific principles of his daily calling would require."—*The Miller.*
"A boon to those who are striving to become efficient mechanics."—*Daily Chronicle.*

MARINE ENGINEERING, SHIPBUILDING, NAVIGATION, etc.

Pocket-Book for Naval Architects and Shipbuilders.

NAVAL ARCHITECT'S & SHIPBUILDER'S POCKET-BOOK

of Formulæ, Rules, and Tables, and Marine Engineer's and Surveyor's Handy Book of Reference. By CLEMENT MACKROW, M.I.N.A. Sixth Edition, Revised, 700 pages, with 300 Illustrations. Fcap., 12s. 6d. leather.

SUMMARY OF CONTENTS.

SIGNS AND SYMBOLS, DECIMAL FRACTIONS.—TRIGONOMETRY.—PRACTICAL GEOMETRY.—MENSURATION.—CENTRES AND MOMENTS OF FIGURES.—MOMENTS OF INERTIA AND RADII OF GYRATION.—ALGEBRAICAL EXPRESSIONS FOR SIMPSON'S RULES.—MECHANICAL PRINCIPLES.—CENTRE OF GRAVITY.—LAWS OF MOTION.—DISPLACEMENT, CENTRE OF BUOYANCY.—CENTRE OF GRAVITY OF SHIP'S HULL.—STABILITY CURVES AND METACENTRES.—SEA AND SHALLOW-WATER WAVES.—ROLLING OF SHIPS.—PROPULSION AND RESISTANCE OF VESSELS.—SPEED TRIALS.—SAILING, CENTRE OF EFFORT.—DISTANCES DOWN RIVERS, COAST LINES.—STEERING AND RUDDERS OF VESSELS.—LAUNCHING CALCULATIONS AND VELOCITIES.—WEIGHT OF MATERIAL AND GEAR.—GUN PARTICULARS AND WEIGHT.—STANDARD GAUGES.—RIVETED JOINTS AND RIVETING.—STRENGTH AND TESTS OF MATERIALS.—BINDING AND SHEARING STRESSES, ETC.—STRENGTH OF SHAFTING, PILLARS, WHEELS, ETC.—HYDRAULIC DATA, ETC.—CONIC SECTIONS, CATENARIAN CURVES.—MECHANICAL POWERS, WORK.—BOARD OF TRADE REGULATIONS FOR BOILERS AND ENGINES.—BOARD OF TRADE REGULATIONS FOR SHIPS.—LLOYD'S RULES FOR BOILERS.—LLOYD'S WEIGHT OF CHAINS.—LLOYD'S SCANTLINGS FOR SHIPS.—DATA OF ENGINES AND VESSELS.—SHIPS' FITTINGS AND TESTS.—SEASONING PRESERVING TIMBER—MEASUREMENT OF TIMBER.—ALLOYS, PAINTS, VARNISHES.—DATA FOR STOWAGE.—ADMIRALTY TRANSPORT REGULATIONS.—RULES FOR HORSE-POWER, SCREW PROPELLERS, ETC.—PERCENTAGES FOR BUTT STRAPS, ETC.—PARTICULARS OF YACHTS.—MASTING AND RIGGING VESSELS.—DISTANCES OF FOREIGN PORTS.—TONNAGE TABLES—VOCABULARY OF FRENCH AND ENGLISH TERMS.—ENGLISH WEIGHTS AND MEASURES.—FOREIGN WEIGHTS AND MEASURES.—DECIMAL EQUIVALENTS.—FOREIGN MONEY.—DISCOUNT AND WAGE TABLES.—USEFUL NUMBERS AND READY RECKONERS.—TABLES OF CIRCULAR MEASURES.—TABLES OF AREAS OF AND CIRCUMFERENCES OF CIRCLES.—TABLES OF AREAS OF SEGMENTS OF CIRCLES.—TABLES OF SQUARES AND CUBES AND ROOTS OF NUMBERS.—TABLES OF LOGARITHMS OF NUMBERS.—TABLES OF HYPERBOLIC LOGARITHMS.—TABLES OF NATURAL SINES, TANGENTS, ETC.—TABLES OF LOGARITHMIC SINES, TANGENTS, ETC.

" In these days of advanced knowledge a work like this is of the greatest value. It contains a vast amount of information. We unhesitatingly say that it is the most valuable compilation for its specific purpose that has ever been printed. No naval architect, engineer, surveyor, or seaman, wood or iron shipbuilder, can afford to be without this work."—*Nautical Magazine.*

" Should be used by all who are engaged in the construction or design of vessels. . . . Will be found to contain the most useful tables and formulæ required by shipbuilders, carefully collected from the best authorities, and put together in a popular and simple form. The book is one of exceptional merit."—*Eng:neer.*

" The professional shipbuilder has now, in a convenient and accessible form, reliable data for solving many of the numerous problems that present themselves in the course of his work."—*Iron.*

" There is no doubt that a pocket-book of this description must be a necessity in the shipbuilding trade. . . . The volume contains a mass of useful information clearly expressed and presented in a handy form."—*Marine Engineer.*

Marine Engineering.

MARINE ENGINES AND STEAM VESSELS: A Treatise on.

By ROBERT MURRAY, C.E. Eighth Edition, thoroughly Revised, with considerable Additions by the Author and by GEORGE CARLISLE, C.E., Senior Surveyor to the Board of Trade at Liverpool. 12mo, 4s. 6d. cloth.

" Well adapted to give the young steamship engineer or marine engine and boiler maker a general introduction into his practical work."—*Mechanical World.*

" We feel sure that this thoroughly revised edition will continue to be as popular in the future as it has been in the past, as, for its size, it contains more useful information than any similar treatise."—*Industries.*

" The information given is both sound and sensible, and well qualified to direct young sea-going hands on the straight road to the extra chief's certificate. . . . Most useful to surveyors, inspectors, draughtsmen, and all young engineers who take an interest in their profession."—*Glasgow Herald.*

English-French Dictionary of Sea Terms.

SEA TERMS, PHRASES, AND WORDS (Technical Dictionary of) used in the English and French Languages. (English-French, French-English). For the Use of Seamen, Engineers, Pilots, Ship-builders, Shipowners and Ship-brokers. Compiled by W. PIRRIE, late of the African Steamship Company. Fcap. 8vo, 5s. cloth limp. [*Just published.*

" This volume will be highly appreciated by seamen, engineers, pilots, shipbuilders and shipowners. It will be found wonderfully accurate and complete "—*Scotsman.*

" A very useful dictionary, which has long been wanted by French and English engineers, masters, officers and others."—*Shipping World.*

Electric Lighting of Ships.

ELECTRIC SHIP LIGHTING: A Handbook on the Practical Fitting and Running of Ship's Electrical Plant, for the Use of Shipowners and Builders, Marine Electricians and Sea-going Engineers in Charge. By J. W. URQUHART, Author of "Electric Light," "Dynamo Construction," &c. With numerous Illustrations. Crown 8vo, 7s. 6d. cloth.

Pocket-Book for Marine Engineers.

MARINE ENGINEERS' POCKET-BOOK of useful Tables and Formulæ. By FRANK PROCTOR, A.I.N.A. Third Edition. Royal 32mo, leather, gilt edges, with strap, 4s.

"We recommend it to our readers as going far to supply a long-felt want."—*Naval Science.*
"A most useful companion to all marine engineers."—*United Service Gazette.*

Introduction to Marine Engineering.

ELEMENTARY ENGINEERING: A Manual for Young Marine Engineers and Apprentices. In the Form of Questions and Answers on Metals, Alloys, Strength of Materials, Construction and Management of Marine Engines and Boilers, Geometry, &c. &c. With an Appendix of Useful Tables. By JOHN SHERREN BREWER, Government Marine Surveyor, Hongkong. Third Edition, small crown 8vo, 1s. 6d. cloth.

"Contains much valuable information for the class for whom it is intended, especially in the chapters on the management of boilers and engines."—*Nautical Magazine.*
"A useful introduction to the more elaborate text books."—*Scotsman.*
"To a student who has the requisite desire and resolve to attain a thorough knowledge, Mr. Brewer offers decidedly useful help."—*Athenæum.*

Navigation.

PRACTICAL NAVIGATION. Consisting of THE SAILOR'S SEA-BOOK, by JAMES GREENWOOD and W. H. ROSSER; together with the requisite Mathematical and Nautical Tables for the Working of the Problems, by HENRY LAW, C.E., and Professor J. R. YOUNG. Illustrated. 12mo, 7s. strongly half-bound.

Drawing for Marine Engineers.

LOCKIE'S MARINE ENGINEER'S DRAWING - BOOK. Adapted to the Requirements of the Board of Trade Examinations. By JOHN LOCKIE, C.E. With 22 Plates, Drawn to Scale. Royal 8vo, 3s. 6d. cloth.

"The student who learns from these drawings will have nothing to unlearn."—*Engineer.*
"The examples chosen are essentially practical, and are such as should prove of service to engineers generally, while admirably fulfilling their specific purpose."—*Mechanical World.*

Sailmaking.

THE ART AND SCIENCE OF SAILMAKING. By SAMUEL B. SADLER, Practical Sailmaker, late in the employment of Messrs. Ratsey and Lapthorne, of Cowes and Gosport. With Plates and other Illustrations. Small 4to, 12s. 6d. cloth.

"This extremely practical work gives a complete education in all the branches of the manufacture, cutting out, roping, seaming, and goring. It is copiously illustrated, and will form a first-rate text-book and guide."—*Portsmouth Times.*
"The author of this work has rendered a distinct service to all interested in the art of sailmaking. The subject of which he treats is a congenial one. Mr. Sadler is a practical sailmaker, and has devoted years of careful observation and study to the subject; and the results of the experience thus gained he has set forth in the volume before us."—*Steamship.*

Chain Cables.

CHAIN CABLES AND CHAINS. Comprising Sizes and Curves of Links, Studs, &c., Iron for Cables and Chains, Chain Cable and Chain Making, Forming and Welding Links, Strength of Cables and Chains, Certificates for Cables, Marking Cables, Prices of Chain Cables and Chains, Historical Notes, Acts of Parliament, Statutory Tests, Charges for Testing, List of Manufacturers of Cables, &c. &c. By THOMAS W. TRAILL, F.E.R.N., M.Inst.C.E., Engineer-Surveyor-in-Chief, Board of Trade, Inspector of Chain Cable and Anchor Proving Establishments, and General Superintendent, Lloyd's Committee on Proving Establishments. With numerous Tables, Illustrations, and Lithographic Drawings. Folio, £2 2s. cloth, bevelled boards.

"It contains a vast amount of valuable information. Nothing seems to be wanting to make it a complete and standard work of reference on the subject."—*Nautical Magazine.*

MINING AND METALLURGY.

Colliery Management, &c.

COLLIERY WORKING AND MANAGEMENT: Comprising the Duties of a Colliery Manager, the Oversight and Arrangement of Labour and Wages, and the different Systems of Working Coal Seams. By II. F. BULMAN and R. A. S. REDMAYNE. 350 pages, with 28 Plates and other Illustrations, including Underground Photographs. Medium 8vo, 15s. cloth. [*Just published.*

" This is, indeed, an admirable Handbook for Colliery Managers, in fact, it is an indispensable adjunct to a Colliery Manager's education, as well as being a most useful and interesting work on the subject for all who in any way have to do wi h coal mining. The underground photographs are an attractive feature of the woi k, being very life-like and necessarily true representations of the scenes they depict."—*Colliery Guardian.*

" Mr. Bulman and Mr. Redmayne, who are both experienced Colliery Managers of great literary ability are to be congratu'ated on having supplied an authoritative work dealing with a side of the subject of coal mining which has hitherto received but scant treatment. The authors elucidate their text by 119 woodcuts and 28 plates, most of the latter being admirable reproductions of photographs taken underground with the aid of the magnesium flash-light. These illustrations are excellent."—*Nature.*

Inflammable Gases.

INFLAMMABLE GAS AND VAPOUR IN THE AIR (the Detection and Measurement of). By FRANK CLOWES, D.Sc., Lond., F.I.C., Prof. of Chemistry in the University College, Nottingham. With a Chapter on THE DETECTION AND MEASUREMENT OF PETROLEUM VAPOUR by BOVERTON REDWOOD, F.R.S.E., Consulting Adviser, to the Corporation of London under the Petroleum Acts. Crown 8vo, cloth, 5s. net. [*Just published.*

" Professor Clowes has given us a volume on a subject of much industrial importance. . . . those interested in these matters may be recommended to study this book, which is easy of comprehension and contains many good things."—*The Engineer.*

"A convenient summary of the work on which Professor Clowes has been engaged for some considerable time. . . It is hardly necessary to say that any work on these subjects with these names on the title-page must be a valuable one, and one that no mining engineer—certainly no coal miner—can afford to ignore or to leave unread."—*Mining Journal.*

Mining Machinery.

MACHINERY FOR METALLIFEROUS MINES: A Practical Treatise for Mining Engineers, Metallurgists and Managers of Mines. By E. HENRY DAVIES, M.E., F.G.S. Crown 8vo, 580 pp., with upwards of 300 Illustrations. 12s. 6d. cloth. [*Just published.*

" Mr. Davies, in this handsome volume, has done the advanced student and the manager of mines good service. Almost every kind of machinery in actual use is carefully described, and the woodcuts and plates are good."—*Athenæum.*

" From cover to cover the work exhibits all the same characteristics which excite the confidence and attract the attention of the student as he peruses the first page. The work may safely be recommended. By its publication the literature connected with the industry will be enriched, and the reputation of its author enhanced."—*Mining Journal.*

Metalliferous Minerals and Mining.

METALLIFEROUS MINERALS AND MINING. By D. C. DAVIES, F.G.S., Mining Engineer, &c., Author of " A Treatise on Slate and Slate Quarrying." Fifth Edition, thoroughly Revised and much Enlarged by his Son, E. HENRY DAVIES, M.E., F.G.S. With about 150 Illustrations. Crown 8vo, 12s. 6d. cloth.

" Neither the practical miner nor the general reader, interested in mines, can have a better book for his companion and his guide."—*Mining Journal.*

" We are doing our readers a service in calling their attention to this valuable work."—*Mining World.*

" As a history of the present state of mining throughout the world this book has a real value, and it supplies an actual want."—*Athenæum.*

Earthy Minerals and Mining.

EARTHY AND OTHER MINERALS AND MINING. By D. C. DAVIES, F.G.S., Author of "Metalliferous Minerals," &c. Third Edition, Revised and Enlarged, by his Son, E. HENRY DAVIES, M.E., F.G.S. With about 100 Illustrations. Crown 8vo, 12s. 6d. cloth.

" We do not remember to have met with any English work on mining matters that contains the same amount of information packed in equally convenient form."—*Academy.*

" We should be inclined to rank it as among the very best of the handy technical and trades manuals which have recently appeared."—*British Quarterly Review.*

Metalliferous Mining in the United Kingdom.

BRITISH MINING : A Treatise on the History, Discovery, Practical Development, and Future Prospects of Metalliferous Mines in the United Kingdom. By ROBERT HUNT, F.R.S., late Keeper of Mining Records. Upwards of 950 pp., with 230 Illustrations. Second Edition, Revised. Super-royal 8vo, £2 2s. cloth.

"The book is a treasure-house of statistical information on mining subjects, and we know of no other work embodying so great a mass of matter of this kind. Were this the only merit of Mr. Hunt's volume it would be sufficient to render it indispensable in the library of everyone interested in the development of the mining and metallurgical industries of this country."—*Athenæum*.

"A mass of information not elsewhere available, and of the greatest value to those who may be interested in our great mineral industries."—*Engineer*.

Underground Pumping Machinery.

MINE DRAINAGE : Being a Complete and Practical Treatise on Direct-Acting Underground Steam Pumping Machinery, with a Description of a large number of the best known Engines, their General Utility and the Special Sphere of their Action, the Mode of their Application, and their merits compared with other forms of Pumping Machinery. By STEPHEN MICHELL. 8vo, 15s. cloth.

"Will be highly esteemed by colliery owners and lessees, mining engineers, and students generally who require to be acquainted with the best means of securing the drainage of mines. It is a most valuable work, and stands almost alone in the literature of steam pumping machinery."—*Colliery Guardian*.

"Much valuable information is given, so that the book is thoroughly worthy of an extensive circulation amongst practical men and purchasers of machinery."—*Mining Journal*.

Prospecting for Gold and other Metals.

THE PROSPECTOR'S HANDBOOK : A Guide for the Prospector and Traveller in Search of Metal-Bearing or other Valuable Minerals. By J. W. ANDERSON, M.A. (Camb.), F.R.G.S., Author of "Fiji and New Caledonia." Sixth Edition, thoroughly Revised and much Enlarged. Small crown 8vo, 3s. 6d. cloth ; or, 4s. 6d. leather, pocket-book form, with tuck.

"Will supply a much felt want, especially among Colonists, in whose way are so often thrown many mineralogical specimens the value of which it is difficult to determine."—*Engineer*.

"How to find commercial minerals, and how to identify them when they are found, are the leading points to which attention is directed. The author has managed to pack as much practical detail into his pages as would supply material for a book three times its size."—*Mining Journal*.

Mining Notes and Formulæ.

NOTES AND FORMULÆ FOR MINING STUDENTS. By JOHN HERMAN MERIVALE, M.A., Certificated Colliery Manager, Professor of Mining in the Durham College of Science, Newcastle-upon-Tyne. Third Edition, Revised and Enlarged. Small crown 8vo, 2s. 6d. cloth.

"The author has done his work in a creditable manner, and has produced a book that will be of service to students, and those who are practically engaged in mining operations."—*Engineer*.

Handybook for Miners.

THE MINER'S HANDBOOK : A Handy Book of Reference on the subjects of Mineral Deposits, Mining Operations, Ore Dressing, &c. For the Use of Students and others interested in Mining matters. Compiled by JOHN MILNE, F.R.S., Professor of Mining in the Imperial University of Japan. Revised Edition. Fcap. 8vo, 7s. 6d. leather.

"Professor Milne's handbook is sure to be received with favour by all connected with mining, and will be extremely popular among students."—*Athenæum*.

Miners' and Metallurgists' Pocket-Book.

A POCKET-BOOK FOR MINERS AND METALLURGISTS. Comprising Rules, Formulæ, Tables, and Notes, for Use in Field and Office Work. By F. DANVERS POWER, F.G.S., M.E. Fcap. 8vo, 9s. leather.

"This excellent book is an admirable example of its kind, and ought to find a large sale amongst English-speaking prospectors and mining engineers."—*Engineering*.

Mineral Surveying and Valuing.

MINERAL SURVEYOR and VALUER'S COMPLETE GUIDE Comprising a Treatise on Improved Mining Surveying and the Valuation of Mining Properties, with New Traverse Tables. By WM. LINTERN. Third Edition, Enlarged. 12mo, 3s. 6d cloth.

Colliery Management.

THE COLLIERY MANAGER'S HANDBOOK: A Comprehensive Treatise on the Laying-out and Working of Collieries, Designed as a Book of Reference for Colliery Managers, and for the Use of Coal-Mining Students preparing for First-class Certificates. By CALEB PAMELY, Mining Engineer and Surveyor; Member of the North of England Institute of Mining and Mechanical Engineers; and Member of the South Wales Institute of Mining Engineers. With nearly 700 Plans, Diagrams, and other Illustrations. Third Edition, Revised and Enlarged, medium 8vo, about 900 pp. Price £1 5s. strongly bound.

SUMMARY OF CONTENTS.

GEOLOGY.—SEARCH FOR COAL.—MINERAL LEASES AND OTHER HOLDINGS.—SHAFT SINKING.—FITTING UP THE SHAFT AND SURFACE ARRANGEMENTS.—STEAM BOILERS AND THEIR FITTINGS.—TIMBERING AND WALLING.—NARROW WORK AND METHODS OF WORKING.—UNDERGROUND CONVEYANCE.—DRAINAGE.—THE GASES MET WITH IN MINES; VENTILATION.—ON THE FRICTION OF AIR IN MINES.—THE PRIESTMAN OIL ENGINE; PETROLEUM AND NATURAL GAS.—SURVEYING AND PLANNING.—SAFETY LAMPS AND FIRE-DAMP DETECTORS.—SUNDRY AND INCIDENTAL OPERATIONS AND APPLIANCES.—COLLIERY EXPLOSIONS.—MISCELLANEOUS QUESTIONS AND ANSWERS.—*Appendix :* SUMMARY OF REPORT OF H.M. COMMISSIONERS ON ACCIDENTS IN MINES.

"Mr. Pamely has not only given us a comprehensive reference book of a very high order, suitable to the requirements of mining engineers and colliery managers, but has also provided mining students with a class-book that is as interesting as it is instructive."—*Colliery Manager.*

"Mr. Pamely's work is eminently suited to the purpose for which it is intended—being clear, interesting, exhaustive, rich in detail, and up to date, giving descriptions of the latest machines in every department. A mining engineer could scarcely go wrong who followed this work."—*Colliery Guardian.*

"This is the most complete ' all-round ' work on coal-mining published in the English language. . . . No library of coal-mining books is complete without it."—*Colliery Engineer* (Scranton, Pa., U.S.A.).

Coal and Iron.

COAL & IRON INDUSTRIES OF THE UNITED KINGDOM. Comprising a Description of the Coal Fields, and of the Principal Seams of Coal, with Returns of their Produce and its Distribution, and Analyses of Special Varieties. Also, an Account of the occurrence of Iron Ores in Veins or Seams ; Analyses of each Variety ; and a History of the Rise and Progress of Pig Iron Manufacture. By RICHARD MEADE. 8vo, £1 8s. cloth.

"Of this book we may unreservedly say that it is the best of its class which we have ever met. . . . A book of reference which no one engaged in the iron or coal trades should omit from his library."—*Iron and Coal Trades' Review.*

Coal Mining.

COAL AND COAL MINING, A Rudimentary Treatise on. By the late Sir WARINGTON W. SMYTH, M.A., F.R.S., &c., Chief Inspector of the Mines of the Crown. Seventh Edition, Revised and Enlarged. With numerous Illustrations, 12mo, 3s. 6d. cloth.

"As an outline is given of every known coal-field in this and other countries, as well as of the principal methods of working, the book will doubtless interest a very large number of readers."—*Mining Journal.*

Asbestos and its Uses.

ASBESTOS: Its Properties, Occurrence, and Uses. With some Account of the Mines of Italy and Canada. By ROBERT H. JONES. With Eight Collotype Plates and other Illustrations. Crown 8vo, 12s. 6d. cloth.

"An interesting and invaluable work."—*Colliery Guardian.*

Subterraneous Surveying.

SUBTERRANEOUS SURVEYING, Elementary and Practical Treatise on ; with and without the Magnetic Needle. By THOMAS FENWICK, Surveyor of Mines, and THOMAS BAKER, C.E. Illustrated. 12mo, 2s. 6d. cloth.

Granite Quarrying.

GRANITES AND OUR GRANITE INDUSTRIES. By GEORGE F. HARRIS, F.G.S., Membre de la Société Belge de Géologie, Lecturer on Economic Geology at the Birkbeck Institution, &c. With Illustrations. Crown 8vo, 2s. 6d. cloth.

"A clearly and well-written manual for persons engaged or interested in the granite industry."—*Scotsman.*

Gold, Metallurgy of.

THE METALLURGY OF GOLD: A Practical Treatise on the Metallurgical Treatment of Gold-bearing Ores. Including the Processes of Concentration, Chlorination, and Extraction by Cyanide, and the Assaying, Melting, and Refining of Gold. By M. EISSLER, Mining Engineer and Metallurgical Chemist, formerly Assistant Assayer of the U.S. Mint, San Francisco. Fourth Edition, Enlarged. With about 250 Illustrations and numerous Folding Plates and Working Drawings. Large crown 8vo, 16s. cloth. *[Just published.*

"This book thoroughly deserves its title of a 'Practical Treatise.' The whole process of gold milling, from the breaking of the quartz to the assay of the bullion, is described in clear and orderly narrative and with much, but not too much, fulness of detail."—*Saturday Review.*

"The work is a storehouse of information and valuable data, and we strongly recommend it to all professional men engaged in the gold-mining industry."—*Mining Journal.*

Gold Extraction.

THE CYANIDE PROCESS OF GOLD EXTRACTION; and its Practical Application on the Witwatersrand Gold Fields in South Africa. By M. EISSLER, M.E., Author of "The Metallurgy of Gold," &c. With Diagrams and Working Drawings. Large crown 8vo, 7s. 6d cloth. *[Just published.*

"This book is just what was needed to acquaint mining men with the actual working of a process which is not only the most popular, but is, as a general rule, the most successful for the extraction of gold from tailings."—*Mining Journal.*

"The work will prove invaluable to all interested in gold mining, whether metallurgists or as investors."—*Chemical News.*

Silver, Metallurgy of.

THE METALLURGY OF SILVER: A Practical Treatise on the Amalgamation, Roasting, and Lixiviation of Silver Ores. Including the Assaying, Melting, and Refining of Silver Bullion. By M. EISSLER, Author of "The Metallurgy of Gold," &c. Third Edition. Crown 8vo, 10s. 6d. cloth.

"A practical treatise, and a technical work which we are convinced will supply a long felt want amongst practical men, and at the same time be of value to students and others indirectly connected with the industries."—*Mining Journal.*

"From first to last the book is thoroughly sound and reliable."—*Colliery Guardian.*

"For chemists, practical miners, assayers, and investors alike, we do not know of any work on the subject so handy and yet so comprehensive."—*Glasgow Herald.*

Lead, Metallurgy of.

THE METALLURGY OF ARGENTIFEROUS LEAD: A Practical Treatise on the Smelting of Silver-Lead Ores and the Refining of Lead Bullion. Including Reports on various Smelting Establishments and Descriptions of Modern Smelting Furnaces and Plants in Europe and America. By M. EISSLER, M.E., Author of "The Metallurgy of Gold," &c. Crown 8vo, 400 pp., with 183 Illustrations, 12s. 6d. cloth.

"The numerous metallurgical processes, which are fully and extensively treated of, embrace all the stages experienced in the passage of the lead from the various natural states to its issue from the refinery as an article of commerce."—*Practical Engineer.*

"The present volume fully maintains the reputation of the author. Those who wish to obtain a thorough insight into the present state of this industry cannot do better than read this volume, and all mining engineers cannot fail to find many useful hints and suggestions in it."—*Industries.*

Iron, Metallurgy of.

METALLURGY OF IRON. By H. BAUERMAN, F.G.S., A.R.S.M. Sixth Edition, Revised and Enlarged. 12mo, 5s. cloth.

Iron Mining.

THE IRON ORES OF GREAT BRITAIN AND IRELAND: Their Mode of Occurrence, Age and Origin, and the Methods of Searching for and Working Them. With a Notice of some of the Iron Ores of Spain. By J. D. KENDALL, F.G.S, Mining Engineer. Crown 8vo, 16s. cloth.

"The author has a thorough practical knowledge of his subject, and has supplemented a careful study of the available literature by unpublished information derived from his own observations. The result is a very useful volume which cannot fail to be of value to all interested in the iron industry of the country."—*Industries.*

ELECTRICITY, ELECTRICAL ENGINEERING, ETC.

Submarine Telegraphy.

SUBMARINE TELEGRAPHS: Their History, Construction and Working. By CHARLES BRIGHT, F.R.S.E. Super royal 8vo, about 600 pages, fully illustrated. [*In the press.*

[*To Subscribers before publication £2 2s. net; after publication the price will be raised to £3 3s. net.*]

Electrical Engineering.

THE ELECTRICAL ENGINEER'S POCKET-BOOK of Modern Rules, Formulæ, Tables, and Data. By H. R. KEMPE, M. Inst. E.E., A.M. Inst. C.E., Technical Officer, Postal Telegraphs, Author of "A Handbook of Electrical Testing," "The Engineer's Year-Book," &c. Second Edition, Thoroughly Revised, with Additions. With numerous Illustrations. Royal 32mo, oblong, 5s. leather.

" There is very little in the shape of formulæ or data which the electrician is likely to want in a hurry which cannot be found in its pages."—*Practical Engineer.*
" A very useful book of reference for daily use in practical electrical engineering and its various applications to the industries of the present day."—*Iron.*
" It is the best book of its kind."—*Electrical Engineer.*
" The Electrical Engineer's Pocket-Book is a good one." - *Electrician.*
" Strongly recommended to those engaged in the electrical industries."— *Electrical Review.*

Electric Lighting.

ELECTRIC LIGHT FITTING: A Handbook for Working Electrical Engineers, embodying Practical Notes on Installation Management. By J. W. URQUHART, Electrician, Author of "Electric Light," &c. With numerous Illus:s. Second Edition, Revised, with Additional Chapters. Crown 8vo, 5s. cloth.

" This volume deals with what may be termed the mechanics of electric lighting, and is addressed to men who are already engaged in the work, or are training for it. The work traverses a great deal of ground, and may be read as a sequel to the same author's useful work on ' Electric Light.'"—*Electrician.*
" This is an attempt to state in the simplest language the precautions which should be adopted in installing the electric light, and to give information for the guidance of those who have to run the plant when installed. The book is well worth the perusal of the workman, for whom it is written."—*Electrical Review.*
" Eminently practical and useful, ... Ought to be in the hands of everyone in charge of an electric light plant."—*Electrical Engineer.*

Electric Light.

ELECTRIC LIGHT: Its Production and Use, Embodying Plain Directions for the Treatment of Dynamo-Electric Machines, Batteries, Accumulators, and Electric Lamps. By J. W. URQUHART, C.E., Author of "Electric Light Fitting," "Electroplating," &c. Fifth Edition, carefully Revised, with Large Additions and 145 Illustrations. Crown 8vo, 7s. 6d. cloth.

" The whole ground of electric lighting is more or less covered and explained in a very clear and concise manner."—*Electrical Review.*
" Contains a good deal of very interesting information, especially in the parts where the author gives dimensions and working costs."—*Electrical Engineer.*
" A vade-mecum of the salient facts connected with the science of electric lighting."—*Electrician.*
" You cannot for your purpose have a better book than ' Electric Light,' by Urquhart."—*Engineer.*
" The book is by far the best that we have yet met with on the subject."—*Athenæum.*

Construction of Dynamos.

DYNAMO CONSTRUCTION: A Practical Handbook for the Use of Engineer Constructors and Electricians-in-Charge. Embracing Framework Building, Field Magnet and Armature Winding and Grouping, Compounding, &c. With Examples of leading English, American, and Continental Dynamos and Motors. By J. W. URQUHART, Author of "Electric Light," &c. Second Edition, Enlarged. With 114 Illustrations. Crown 8vo, 7s. 6d. cloth.

" Mr. Urquhart's book is the first one which deals with these matters in such a way that the engineering student can understand them. The book is very readable, and the author leads his readers up to difficult subjects by reasonably simple tests."—*Engineering Review.*
" ' Dynamo Construction' more than sustains the high character of the author's previous publications. It is sure to be widely read by the large and rapidly-increasing number of practical electricians."—*Glasgow Herald.*
" A book for which a demand has long existed."—*Mechanical World.*

Dynamo Management.

THE MANAGEMENT OF DYNAMOS: A Handybook of Theory and Practice for the Use of Mechanics, Engineers, Students and others in Charge of Dynamos. By G. W. LUMMIS PATERSON. With numerous Illustrations. Crown 8vo, 3s. 6d. cloth.

"An example which deserves to be taken as a model by other authors. The subject is treated in a manner which any intelligent man who is fit to be entrusted with charge of an engine should be able to understand. It is a useful book to all who make, tend or employ electric machinery."—*Architect.*

A New Dictionary of Electricity.

THE STANDARD ELECTRICAL DICTIONARY. A Popular Dictionary of Words and Terms Used in the Practice of Electrical Engineering. Containing upwards of 3,000 Definitions. By T. O'CONOR SLOANE, A.M., Ph.D. Crown 8vo, 630 pp., 350 Illustrations, 7s. 6d. cloth.

"The work has many attractive features in it, and is, beyond doubt, a well put together and useful publication. The amount of ground covered may be gathered from the fact that in the index about 5,000 references will be found."—*Electrical Review.*

Electric Lighting of Ships.

ELECTRIC SHIP-LIGHTING: A Handbook on the Practical Fitting and Running of Ship's Electrical Plant. For the Use of Shipowners and Builders, Marine Electricians, and Sea-going Engineers in Charge. By J. W. URQUHART, C.E. With 88 Illustrations, crown 8vo, 7s. 6d. cloth.

"The subject of ship electric lighting is one of vast importance, and Mr. Urquhart is to be highly complimented for placing such a valuable work at the service of marine electricians."—*The Steamship*

Country House Electric Lighting.

ELECTRIC LIGHT FOR COUNTRY HOUSES: A Practical Handbook on the Erection and Running of Small Installations, with Particulars of the Cost of Plant and Working. By J. H. KNIGHT. Crown 8vo, 1s. wrapper.

"The book contains excellent advice and many practical hints for the help of those who wish to light their own houses."—*Building News.*

Electric Lighting.

ELEMENTARY PRINCIPLES OF ELECTRIC LIGHTING. By ALAN A. CAMPBELL SWINTON, Associate I.E.E. Third Edition, Enlarged and Revised. With Sixteen Illustrations. Crown 8vo, 1s. 6d. cloth.

"Anyone who desires a short and thoroughly clear exposition of the elementary principles of electric-lighting cannot do better than read this little work."—*Bradford Observer.*

Dynamic Electricity.

ELEMENTS OF DYNAMIC ELECTRICITY & MAGNETISM. By PHILIP ATKINSON, A.M., Ph.D., Author of "Elements of Static Electricity,' &c. Crown 8vo, 417 pp., with 120 Illustrations, 10s. 6d. cloth.

Electric Motors, &c.

THE ELECTRIC TRANSFORMATION OF POWER and its Application by the Electric Motor, including Electric Railway Construction. By P. ATKINSON, A.M., Ph.D. With 96 Illustrations. Crown 8vo, 7s. 6d. cloth.

Dynamo Construction.

HOW TO MAKE A DYNAMO: A Practical Treatise for Amateurs. Containing numerous Illustrations and Detailed Instructions for Constructing a Small Dynamo to Produce the Electric Light. By ALFRED CROFTS. Fifth Edition, Revised and Enlarged. Crown 8vo, 2s. cloth.   [*Just published.*

"The instructions given in this unpretentious little book are sufficiently clear and explicit to enable any amateur mechanic possessed of average skill and the usual tools to be found in an amateur's workshop, to build a practical dynamo machine."—*Electrician.*

Text-Book of Electricity.

THE STUDENT'S TEXT-BOOK OF ELECTRICITY. By H. M. NOAD, F.R.S. Cheaper Edition. 650 pp., with 470 Illustrations. Crown 8vo, 9s. cloth.

ARCHITECTURE, BUILDING, etc.

Building Construction.

PRACTICAL BUILDING CONSTRUCTION: A Handbook for Students Preparing for Examinations, and a Book of Reference for Persons Engaged in Building. By JOHN PARNELL ALLEN, Surveyor, Lecturer on Building Construction at the Durham College of Science, Newcastle-on-Tyne. Medium 8vo, 450 pages, with 1,000 Illustrations. 12s. 6d. cloth.

"The most complete exposition of building construction we have seen. It contains all that is necessary to prepare students for the various examinations in building construction."—*Building News.*

"The author depends nearly as much on his diagrams as on his type. The pages suggest the hand of a man of experience in building operations—and the volume must be a blessing to many teachers as well as to students."—*The Architect.*

"The work is sure to prove a formidable rival to great and small competitors alike, and bids fair to take a permanent place as a favourite students' text-book. The large number of illustrations deserve particular mention for the great merit they possess for purposes of reference, in exactly corresponding to convenient scales."—*Jour. Inst. Brit. Archts.*

Masonry.

PRACTICAL MASONRY: A Guide to the Art of Stone Cutting. Comprising the Construction, Setting-Out, and Working of Stairs, Circular Work, Arches, Niches, Domes, Pendentives, Vaults, Tracery Windows, &c. &c. For the Use of Students, Masons, and other Workmen. By WILLIAM R. PURCHASE, Building Inspector to the Town of Hove. Royal 8vo, 134 pages, with 50 Lithographic Plates, comprising nearly 400 separate Diagrams, 7s. 6d. cloth.

"Mr. Purchase's 'Practical Masonry' will undoubtedly be found useful to all interested in this important subject, whether theoretically or practically. Most of the examples given are from actual work carried out, the diagrams being carefully drawn. The book is a practical treatise on the subject, the author himself having commenced as an operative mason, and afterwards acted as foreman mason on many large and important buildings prior to the attainment of his present position. It should be found of general utility to architectural students and others, as well as to those to whom it is specially addressed."—*Journal of the Royal Institute of British Architects.*

"The author has evidently devoted much time and conscientious labour in the production of his book, which will be found very serviceable to students, masons, and other workmen, while its value is much enhanced by the capital illustrations, consisting of fif y lithographic plates, comprising about 400 diagrams."—*Illustrated Carpenter and Builder.*

Concrete.

CONCRETE: ITS NATURE AND USES. A Book for Architects, Builders, Contractors, and Clerks of Works. By GEORGE L. SUTCLIFFE, A.R.I.B.A. 350 pages, with numerous Illustrations. Crown 8vo, 7s. 6d. cloth.

"The author treats a difficult subject in a lucid manner. The manual fills a long felt gap. It is careful and exhaustive; equally useful as a student's guide and an architect's book of reference."—*Journal of Royal Institution of British Architects.*

"There is room for this new book, which will probably be for some time the standard work on the subject for a builder's purpose."—*Glasgow Herald.*

Mechanics for Architects.

THE MECHANICS OF ARCHITECTURE: A Treatise on Applied Mechanics, especially Adapted to the Use of Architects. By E. W. TARN, M.A., Author of "The Science of Building," &c. Second Edition, Enlarged. Illustrated with 125 Diagrams. Crown 8vo, 7s. 6d. cloth.

"The book is a very useful and helpful manual of architectural mechanics, and really contains sufficient to enable a careful and painstaking student to grasp the principles bearing upon the majority of building problems. . . . Mr. Tarn has added, by this volume, to the debt of gratitude which is owing to him by architectural students for the many valuable works which he has produced for their use."—*The Builder.*

"The mechanics in the volume are really mechanics, and are harmoniously wrought in with the distinctive professional matter proper to the subject. The diagrams and type are commendably clear."—*The Schoolmaster.*

The New Builder's Price Book, 1897.

LOCKWOOD'S BUILDER'S PRICE BOOK FOR 1897. A Comprehensive Handbook of the Latest Prices and Data for Builders, Architects, Engineers, and Contractors. Re-constructed, Re-written, and Greatly Enlarged. By FRANCIS T. W. MILLER. 800 closely-printed pages, crown 8vo, 4s. cloth.

"This book is a very useful one, and should find a place in every English office connected with the building and engineering professions."—*Industries.*

"An excellent book of reference."—*Architect.*

"In its new and revised form this Price Book is what a work of this kind should be—comprehensive, reliable, well arranged, legible, and well bound."—*British Architect.*

The New London Building Act.

THE LONDON BUILDING ACT, 1894. With the By-Laws and Regulations of the London County Council, and Introduction, Notes, Cases and Index. By ALEX. J. DAVID, B.A., LL.M., of the Inner Temple, Barrister-at-Law. Crown 8vo, 3s. 6d. cloth.
"To all architects and district surveyors and builders, Mr. David's manual will be welcome."—*Building News.*
"The volume will doubtless be eagerly consulted by the building fraternity."—*Illustrated Carpenter and Builder.*

Designing Buildings.

THE DESIGN OF BUILDINGS: Being Elementary Notes on the Planning, Sanitation and Ornamentive Formation of Structures, based on Modern Practice. Illustrated with Nine Folding Plates. By W. WOODLEY. 8vo, 6s. cloth.

Sir William Chambers's Treatise on Civil Architecture.

THE DECORATIVE PART OF CIVIL ARCHITECTURE. By Sir WILLIAM CHAMBERS, F.R.S. With Portrait, Illustrations, Notes, and an EXAMINATION OF GRECIAN ARCHITECTURE, by JOSEPH GWILT, F.S.A. Revised and Edited by W. H. LEEDS. 66 Plates, 4to, 21s. cloth.

Villa Architecture.

A HANDY BOOK OF VILLA ARCHITECTURE: Being a Series of Designs for Villa Residences in various Styles. With Outline Specifications and Estimates. By C. WICKES, Architect, Author of "The Spires and Towers of England," &c. 61 Plates, 4to, £1 11s. 6d. half-morocco, gilt edges.
"The whole of the designs bear evidence of their being the work of an artistic architect, and they will prove very valuable and suggestive."—*Building News.*

Text-Book for Architects.

THE ARCHITECT'S GUIDE: Being a Text-book of Useful Information for Architects, Engineers, Surveyors, Contractors, Clerks of Works, &c. &c. By FREDERICK ROGERS, Architect. Third Edition. Cr. 8vo, 3s. 6d. cloth.
"As a text-book of useful information for architects, engineers, surveyors, &c., it would be hard to find a handier or more complete little volume."—*Standard.*

Linear Perspective.

ARCHITECTURAL PERSPECTIVE. The whole Course and Operations of the Draughtsman in Drawing a Large House in Linear Perspective. Illustrated by 43 Folding Plates. By F. O. FERGUSON. Second Edition, Enlarged. 8vo, 3s. 6d. boards.
"It is the most intelligible of the treatises on this ill-treated subject that I have met with."—E. INGRESS BELL, ESQ., *in the R.I.B.A. Journal.*

Architectural Drawing.

PRACTICAL RULES ON DRAWING, for the Operative Builder and Young Student in Architecture. By GEORGE PYNE. 14 Plates, 4to, 7s. 6d. bds.

Designing, Measuring, and Valuing.

MEASURING AND VALUING ARTIFICER'S WORK (The Student's Guide to the Practice of). Containing Directions for taking Dimensions, Abstracting the same, and bringing the Quantities into Bill, with Tables of Constants for Valuation of Labour, and for the Calculation of Areas and Solidities. Originally edited by E. DOBSON, Architect. With Additions by E. W. TARN, M.A. Sixth Edition. With 8 Plates and 63 Woodcuts. Crown 8vo, 7s. 6d. cloth.
"This edition will be found the most complete treatise on the principles of measuring and valuing artificers' work that has yet been published."—*Building News.*

Pocket Estimator and Technical Guide.

POCKET TECHNICAL GUIDE, MEASURER, & ESTIMATOR for Builders and Surveyors. Containing Technical Directions for Measuring Work in all the Building Trades, Complete Specifications for Houses, Roads, and Drains, and an Easy Method of Estimating the parts of a Building collectively. By A. C. BEATON. Eighth Edition. Waistcoat-pocket size, 1s. 6d. gilt edges.
"No builder, architect, surveyor, or valuer should be without his 'Beaton.'"—*Building News.*

Constructional Steel Work, &c.

CONSTRUCTIONAL IRON AND STEEL WORK. As Applied to Public, Private and Domestic Buildings. A Practical Treatise for Architects, Students and Builders. By F. CAMPIN. Crown 8vo, 3s. 6d. cloth. [*Just published.*

" Anyone who wants a book on ironwork, as employed in buildings for stanchions, columns and beams, will find the present volume to be suitable. The author has had long and varied experience in designing this class of work. The illustrations have the character of working drawings. This practical book may be counted a most valuable work."—*British Architect.*

Bartholomew and Rogers' Specifications.

SPECIFICATIONS FOR PRACTICAL ARCHITECTURE. A Guide to the Architect, Engineer, Surveyor, and Builder. With an Essay on the Structure and Science of Modern Buildings. Upon the Basis of the Work by ALFRED BARTHOLOMEW, thoroughly Revised, Corrected, and greatly added to by FREDERICK ROGERS, Architect. Third Edition, Revised. 8vo, 15s. cloth.

" The work is too well known to need any recommendation from us. It is one of the books with which every young architect must be equipped."—*Architect.*

Construction.

THE SCIENCE OF BUILDING: An Elementary Treatise on the Principles of Construction. By E. WYNDHAM TARN, M.A., Architect. Third Edition, Revised and Enlarged, with 59 Engravings. Fcap. 8vo, 4s. cloth.

" A very valuable book, which we strongly recommend to all students."—*Builder.*

House Building and Repairing.

THE HOUSE-OWNER'S ESTIMATOR; or, What will it Cost to Build, Alter, or Repair? A Price Book for Unprofessional People, as well as the Architectural Surveyor and Builder. By J. D. SIMON. Edited by F. T. W. MILLER, A.R.I.B.A. Fourth Edition. Crown 8vo, 3s. 6d. cloth.

" In two years it will repay its cost a hundred times over."—*Field.*

Building ; Civil and Ecclesiastical.

A BOOK ON BUILDING, Civil and Ecclesiastical, including Church Restoration ; with the Theory of Domes and the Great Pyramid, &c. By Sir EDMUND BECKETT, Bart., LL.D., F.R.A.S. Second Edit. Fcap. 8vo, 4s. 6d. cloth.

" A book which is always amusing and nearly always instructive."—*The Times.*

Sanitary Houses, etc.

THE SANITARY ARRANGEMENT OF DWELLING HOUSES: A Handbook for Householders and Owners of Houses. By A. J. WALLIS-TAYLER, A.M. Inst. C.E. With numerous Illustrations. Crown 8vo, 2s. 6d. cloth.

" This book will be largely read ; it will be of considerable service to the public. It is well arranged, easily read, and for the most part devoid of technical terms."—*Lancet.*

Ventilation of Buildings.

VENTILATION. A Text-Book to the Practice of the Art of Ventilating Buildings. By W. P. BUCHAN, R.P. 12mo, 4s cloth.

" Contains a great amount of useful practical information, as thoroughly interesting as it is technically reliable.'"—*British Architect.*

The Art of Plumbing.

PLUMBING. A Text-Book to the Practice of the Art or Craft of the Plumber. By W. P. BUCHAN, R.P. Sixth Edition, Enlarged. 12mo, 3s. 6d. cloth.

" A text book which may be safely put in the hands of every young plumber."—*Builder.*

Geometry for the Architect, Engineer, &c.

PRACTICAL GEOMETRY, for the Architect, Engineer, and Mechanic. Giving Rules for the Delineation and Application of various Geometrical Lines, Figures, and Curves. By E. W. TARN, M.A., Architect. 8vo, 9s. cloth.

" No book with the same objects in view has ever been published in which the clearness of the rules laid down and the illustrative diagrams have been so satisfactory."—*Scotsman.*

The Science of Geometry.

THE GEOMETRY OF COMPASSES ; or, Problems Resolved by the mere Description of Circles, and the use of Coloured Diagrams and Symbols. By OLIVER BYRNE. Coloured Plates. Crown 8vo, 3s. 6d. cloth.

CARPENTRY, TIMBER, etc.

Tredgold's Carpentry, Revised and Enlarged by Tarn.

THE ELEMENTARY PRINCIPLES OF CARPENTRY: A Treatise on the Pressure and Equilibrium of Timber Framing, the Resistance of Timber, and the Construction of Floors, Arches, Bridges, Roofs, Uniting Iron and Stone with Timber, &c. To which is added an Essay on the Nature and Properties of Timber, &c., with Descriptions of the kinds of Wood used in Building ; also numerous Tables of the Scantlings of Timber for different purposes, the Specific Gravities of Materials, &c. By THOMAS TREDGOLD, C.E. With an Appendix of Specimens of Various Roofs of Iron and Stone, Illustrated, Seventh Edition, thoroughly Revised and considerably Enlarged by E. WYNDHAM TARN, M.A., Author of "The Science of Building," &c. With 61 Plates, Portrait of the Author, and several Woodcuts. In One large Vol., 4to, 25s. cloth.

"Ought to be in every architect's and every builder's library."—*Builder.*
"A work whose monumental excellence must commend it wherever skilful carpentry is concerned. The author's principles are rather confirmed than impaired by time. The additional plates are of great intrinsic value."—*Building News.*

Woodworking Machinery.

WOODWORKING MACHINERY: Its Rise, Progress, and Con- struction. With Hints on the Management of Saw Mills and the Economical Conversion of Timber. Illustrated with Examples of Recent Designs by leading English, French, and American Engineers. By M. POWIS BALE, A.M.Inst.C.E., M.I.M.E. Second Edition, Revised, with large Additions, large crown 8vo, 440 pp., 9s. cloth.

"Mr. Bale is evidently an expert on the subject, and he has collected so much information that his book is all-sufficient for builders and others engaged in the conversion of timber."—*Architect.*
"The most comprehensive compendium of wood-working machinery we have seen. The author is a thorough master of his subject."—*Building News.*

Saw Mills.

SAW MILLS: Their Arrangement and Management, and the Economical Conversion of Timber. (A Companion Volume to "Woodworking Machinery.") By M. POWIS BALE, A.M.Inst.C.E. Second Edition, Revised. Crown 8vo, 10s. 6d. cloth. [*Just published.*

"The *administration* of a large sawing establishment is discussed, and the subject examined from a financial standpoint. Hence the size, shape, order, and disposition of saw-mills and the like are gone into in detail, and the course of the timber is traced from its reception to its delivery in its converted state. We could not desire a more complete or practical treatise."—*Builder.*

Nicholson's Carpentry.

THE CARPENTER'S NEW GUIDE ; or, Book of Lines for Car- penters ; comprising all the Elementary Principles essential for acquiring a knowledge of Carpentry. Founded on the late PETER NICHOLSON'S standard work. A New Edition, Revised by ARTHUR ASHPITEL, F.S.A. Together with Practical Rules on Drawing, by GEORGE PYNE. With 74 Plates, 4to, £1 1s. cloth.

Handrailing and Stairbuilding.

A PRACTICAL TREATISE ON HANDRAILING : Showing New and Simple Methods for Finding the Pitch of the Plank, Drawing the Moulds, Bevelling, Jointing-up, and Squaring the Wreath. By GEORGE COLLINGS. Second Edition, Revised and Enlarged, to which is added A TREATISE ON STAIR-BUILDING. With Plates and Diagrams. 12mo, 2s. 6d. cloth.

"Will be found of practical utility in the execution of this difficult branch of joinery."—*Builder.*
"Almost every difficult phase of this somewhat intricate branch of joinery is elucidated by the aid of plates and explanatory letterpress."—*Furniture Gazette.*

Circular Work.

CIRCULAR WORK IN CARPENTRY AND JOINERY: A Practical Treatise on Circular Work of Single and Double Curvature. By GEORGE COLLINGS. With Diagrams. Second Edition, 12mo, 2s. 6d. cloth.

"An excellent example of what a book of this kind should be. Cheap in price, clear in definition, and practical in the examples selected."—*Builder.*

Handrailing.

HANDRAILING COMPLETE IN EIGHT LESSONS. On the Square-Cut System. By J. S. GOLDTHORP, Teacher of Geometry and Building Construction at the Halifax Mechanic's Institute. With Eight Plates and over 150 Practical Exercises. 4to, 3*s*. 6*d*. cloth.

"Likely to be of considerable value to joiners and others who take a pride in good work. The arrangement of the book is excellent. We heartily commend it to teachers and students."—*Timber Trades Journal.*

Timber Merchant's Companion.

TIMBER MERCHANT'S and BUILDER'S COMPANION. Containing New and Copious Tables of the Reduced Weight and Measurement of Deals and Battens, of all sizes, from One to a Thousand Pieces, and the relative Price that each size bears per Lineal Foot to any given Price per Petersburgh Standard Hundred; the Price per Cube Foot of Square Timber to any given Price per Load of 50 Feet; the proportionate Value of Deals and Battens by the Standard, to Square Timber by the Load of 50 Feet; the readiest mode of ascertaining the Price of Scantling per Lineal Foot of any size, to any given Figure per Cube Foot, &c. &c. By WILLIAM DOWSING. Fourth Edition, Revised and Corrected. Crown 8vo, 3*s*. cloth.

"Everything is as concise and clear as it can possibly be made. There can be no doubt that every timber merchant and builder ought to possess it."—*Hull Advertiser.*

"We are glad to see a fourth edition of these admirable tables, which for correctness and simplicity of arrangement leave nothing to be desired."—*Timber Trades' Journal.*

Practical Timber Merchant.

THE PRACTICAL TIMBER MERCHANT: Being a Guide for the use of Building Contractors, Surveyors, Builders, &c., comprising useful Tables for all purposes connected with the Timber Trade, Marks of Wood, Essay on the Strength of Timber, Remarks on the Growth of Timber, &c. By W. RICHARDSON. Second Edition. Fcap. 8vo, 3*s*. 6*d*. cloth.

"This handy manual contains much valuable information for the use of timber merchants, builders, foresters, and all others connected with the growth, sale, and manufacture of timber."—*Journal of Forestry.*

Packing-Case Makers, Tables for.

PACKING-CASE TABLES; showing the number of Superficial Feet in Boxes or Packing-Cases, from six inches square and upwards. By W. RICHARDSON, Timber Broker. Third Edition. Oblong 4to, 3*s*. 6*d*. cloth.

"Invaluable labour-saving tables."—*Ironmonger.* "Will save much labour and calculation."—*Grocer*

Superficial Measurement.

TRADESMAN'S GUIDE to SUPERFICIAL MEASUREMENT. Tables calculated from 1 to 200 inches in length, by 1 to 108 inches in breadth. For the use of Architects, Surveyors, Engineers, Timber Merchants, Builders, &c. By JAMES HAWKINGS. Fourth Edition. Fcap., 3*s*. 6*d*. cloth.

"A useful collection of tables to facilitate rapid calculation of surfaces. The exact area of any surface of which the limits have been ascertained can be instantly determined. The book will be found of the greatest utility to all engaged in building operations."—*Scotsman.*

"These tables will be found of great assistance to all who require to make calculations in superficial measurement."—*English Mechanic.*

Forestry.

THE ELEMENTS OF FORESTRY. Designed to afford Information concerning the Planting and Care of Forest Trees for Ornament or Profit, with suggestions upon the Creation and Care of Woodlands. By F. B. HOUGH. Large crown 8vo, 10*s*. cloth.

Timber Importer's Guide.

THE TIMBER IMPORTER'S, TIMBER MERCHANT'S, and Builder's Standard Guide. By RICHARD E. GRANDY. Comprising:—An Analysis of Deal Standards, Home and Foreign, with Comparative Values and Tabular Arrangements for fixing Net Landed Cost on Baltic and North American Deals, including all intermediate Expenses, Freight, Insurance, &c. &c.; together with copious Information for the Retailer and Builder. Third Edition, Revised. 12mo, 2*s*. cloth.

"Everything it pretends to be: built up gradually, it leads one from a forest to a treenail, and throws in as a makeweight, a host of material concerning bricks, columns, cisterns, &c."—*English Mechanic.*

DECORATIVE ARTS, etc.

Woods and Marbles, Imitation of.

SCHOOL OF PAINTING FOR THE IMITATION OF WOODS AND MARBLES, as Taught and Practised by A. R. VAN DER BURG and P. VAN DER BURG, Directors of the Rotterdam Painting Institution. Royal folio, 18½ by 12½ in., Illustrated with 24 full-size Coloured Plates; also 12 plain Plates, comprising 154 Figures. Second and Cheaper Edition. Price £1 11s. 6d.

LIST OF PLATES.

VARIOUS TOOLS REQUIRED FOR WOOD PAINTING.—2, 3. WALNUT; PRELIMINARY STAGES OF GRAINING AND FINISHED SPECIMEN.—4. TOOLS USED FOR MARBLE PAINTING AND METHOD OF MANIPULATION. — 5, 6. ST. REMI MARBLE; EARLIER OPERATIONS AND FINISHED SPECIMEN. —7. METHODS OF SKETCHING DIFFERENT GRAINS, KNOTS, &c.—8, 9. ASH: PRELIMINARY STAGES AND FINISHED SPECIMEN.—10. METHODS OF SKETCHING MARBLE GRAINS.—11, 12. BRECHE MARBLE; PRELIMINARY STAGES OF WORKING AND FINISHED SPECIMEN.—13. MAPLE; ME- THODS OF PRODUCING THE DIFFERENT GRAINS. —14, 15. BIRD'S-EYE MAPLE; PRELIMINARY STAGES AND FINISHED SPECIMEN.—16. METHODS OF SKETCHING THE DIFFERENT SPECIES OF WHITE MARBLE.—17, 18. WHITE MARBLE; PRE- LIMINARY STAGES OF PROCESS AND FINISHED SPECIMEN.—19. MAHOGANY; SPECIMENS OF VARI- OUS GRAINS AND METHODS OF MANIPULATION.— 20, 21. MAHOGANY; EARLIER STAGES AND FINISHED SPECIMEN.—22, 23 24. SIENNA MARBLE; VARIETIES OF GRAIN; PRELIMINARY STAGES AND FINISHED SPECIMEN.—25, 26, 27. JUNIPER WOOD; METHODS OF PRODUCING GRAIN, &c.; PRELIMI- NARY STAGES AND FINISHED SPECIMEN.—28, 29 30. VERT DE MER MARBLE; VARIETIES OF GRAIN AND METHODS OF WORKING, UNFINISHED AND FINISHED SPECIMENS.—31, 32, 33. OAK; VARIETIES OF GRAIN, TOOLS EMPLOYED AND METHODS OF MANIPULATION, PRELIMINARY STAGES AND FINISHED SPECIMEN.—34, 35, 36, WAULSORT MARBLE; VARIETIES OF GRAIN, UNFINISHED AND FINISHED SPECIMENS.

"Those who desire to attain skill in the art of painting woods and marbles will find advantage in consulting this book. . . . Some of the Working Men's Clubs should give their young men the opportunity to study it."—*Builder.*

"A comprehensive guide to the art The explanations of the processes, the manipulation and manage- ment of the colours, and the beautifully executed plates will not be the least valuable to the student who aims at making his work a faithful transcript of nature."—*Building News.*

"Students and novices are fortunate who are able to become the possessors of so noble a work."— *The Architect.*

House Decoration.

ELEMENTARY DECORATION: A Guide to the Simpler Forms of Everyday Art. Together with PRACTICAL HOUSE DECORATION. By JAMES W. FACEY. With numerous Illustrations. In One Vol., 5s. strongly half-bound.

House-Painting, Graining, etc.

HOUSE-PAINTING, GRAINING, MARBLING, AND SIGN WRITING, A Practical Manual of. By ELLIS A. DAVIDSON. Seventh Edition. With Coloured Plates and Wood Engravings. 12mo, 6s. cloth boards.

"A mass of information, of use to the amateur and of value to the practical man."—*English Mechanic.*

Decorators, Receipts for.

THE DECORATOR'S ASSISTANT: A Modern Guide for Decora- tive Artists and Amateurs, Painters, Writers, Gilders, &c. Containing upwards of 600 Receipts, Rules and Instructions; with a variety of Information for General Work connected with every Class of Interior and Exterior Decorations, &c., Sixth Edition. 152 pp., crown 8vo, 1s. in wrapper.

"Full of receipts of value to decorators, painters, gilders, &c. . The book contains the gist of larger treatises on colour and technical processes. It would be difficult to meet with a work so full of varied information on the painter's art."—*Building News.*

British and Foreign Marbles.

MARBLE DECORATION and the Terminology of British and Foreign Marbles. A Handbook for Students. By GEORGE H. BLAGROVE, Author of "Shoring and its Application," &c. With 28 Illustrations. Crown 8vo, 3s. 6d. cloth.

"This most useful and much wanted handbook should be in the hands of every architect and builder."—*Building World.*

"A carefully and usefully written treatise; the work is essentially practical."—*Scotsman.*

DELAMOTTE'S WORKS ON ILLUMINATION AND ALPHABETS.

A PRIMER OF THE ART OF ILLUMINATION, for the Use of
Beginners ; with a Rudimentary Treatise on the Art, Practical Directions for its
Exercise, and Examples taken from Illuminated MSS., printed in Gold and Colours.
By F. DELAMOTTE. New and Cheaper Edition. Small 4to, 6s. ornamental boards.

"The examples of ancient MSS. recommended to the student, which, with much good sense, the author chooses from collections accessible to all, are selected with judgment and knowledge, as well as taste."—*Athenæum.*

ORNAMENTAL ALPHABETS, Ancient and Mediæval, from the Eighth
Century, with Numerals ; including Gothic, Church-Text, large and small, German,
Italian, Arabesque, Initials for Illumination, Monograms, Crosses, &c. &c., for the
use of Architectural and Engineering Draughtsmen, Missal Painters, Masons,
Decorative Painters, Lithographers, Engravers, Carvers, &c. &c. Collected and
Engraved by F. DELAMOTTE, and printed in Colours. New and Cheaper Edition.
Royal 8vo, oblong, 2s. 6d. ornamental boards.

"For those who insert enamelled sentences round gilded chalices, who blazon shop legends over shop-doors, who letter church walls with pithy sentences from the Decalogue, this book will be useful."—*Athenæum.*

EXAMPLES OF MODERN ALPHABETS, Plain and Ornamental,
including German, Old English, Saxon, Italic, Perspective, Greek, Hebrew, Court
Hand, Engrossing, Tuscan, Riband, Gothic, Rustic, and Arabesque ; with several
Original Designs, and an Analysis of the Roman and Old English Alphabets, large
and small, and Numerals, for the use of Draughtsmen, Surveyors, Masons, Decorative Painters, Lithographers, Engravers, Carvers, &c. Collected and Engraved by
F. DELAMOTTE, and printed in Colours. New and Cheaper Edition. Royal 8vo,
oblong, 2s. 6d. ornamental boards.

"There is comprised in it every possible shape into which the letters of the alphabet and numerals can be formed, and the talent which has been expended in the conception of the various plain and ornamental letters is wonderful."—*Standard.*

MEDIÆVAL ALPHABETS & INITIALS FOR ILLUMINATORS.
By F. G. DELAMOTTE. Containing 21 Plates and Illuminated Title, printed in
Gold and Colours. With an Introduction by J. WILLIS BROOKS. Fourth and
Cheaper Edition. Small 4to, 4s. ornamental boards.

"A volume in which the letters of the alphabet come forth glorified in gilding and all the colours of the prism interwoven and intertwined and intermingled."—*Sun.*

THE EMBROIDERER'S BOOK OF DESIGN. Containing Initials,
Emblems, Cyphers, Monograms, Ornamental Borders, Ecclesiastical Devices,
Mediæval and Modern Alphabets, and National Emblems. Collected by F. DELA-
MOTTE, and printed in Colours. Oblong royal 8vo, 1s. 6d. ornamental wrapper.

"The book will be of great assistance to ladies and young children who are endowed with the art of plying the needle in this most ornamental and useful pretty work."—*East Anglian Times.*

Wood Carving.

INSTRUCTIONS IN WOOD-CARVING FOR AMATEURS ;
with Hints on Design. By A LADY. With Ten Plates. New and Cheaper Edition. Crown 8vo, 2s. in emblematic wrapper.

"The handicraft of the wood-carver, so well as a book can impart it, may be learnt from 'A Lady's publication."—*Athenæum.*

Painting, Fine Art.

PAINTING POPULARLY EXPLAINED. By THOMAS JOHN
GULLICK, Painter, and JOHN TIMBS, F.S.A. Including Fresco, Oil, Mosaic,
Water Colour, Water-Glass, Tempera, Encaustic, Miniature, Painting on Ivory,
Vellum, Pottery, Enamel, Glass, &c. Fifth Edition. Crown 8vo, 5s. cloth.

** Adopted as a Prize book at South Kensington.

"Much may be learned, even by those who fancy they do not require to be taught, from the care-ful perusal of this unpretending but comprehensive treatise."—*Art Journal.*

NATURAL SCIENCE, etc.

The Heavens and their Origin.

THE VISIBLE UNIVERSE: Chapters on the Origin and Construction of the Heavens. By J. E. GORE, F.R.A.S., Author of "Star Groups," &c. Illustrated by 6 Stellar Photographs and 12 Plates. Demy 8vo, 16s. cloth.

"A valuable and lucid summary of recent astronomical theory, rendered more valuable and attractive by a series of stellar photographs and other illustrations."—*The Times.*

"In presenting a clear and concise account of the present state of our knowledge, Mr. Gore has made a valuable addition to the literature of the subject."—*Nature.*

"Mr. Gore's 'Visible Universe' is one of the finest works on astronomical science that has recently appeared in our language. In spirit and in method it is scientific from cover to cover, but the style is so clear and attractive that it will be as acceptable and as readable to those who make no scientific pretensions as to those who devote themselves specially to matters astronomical."—*Leeds Mercury.*

The Constellations.

STAR GROUPS: A Student's Guide to the Constellations. By J. ELLARD GORE, F.R.A.S., M.R.I.A., &c., Author of "The Visible Universe," "The Scenery of the Heavens," &c. With 30 Maps. Small 4to, 5s. cloth.

"The volume contains thirty maps showing stars of the sixth magnitude—the usual naked-eye limit—and each is accompanied by a brief commentary, adapted to facilitate recognition and bring to notice objects of special interest. For the purpose of a preliminary survey of the 'midnight pomp' of the heavens, nothing could be better than a set of delineations averaging scarcely twenty square inches in area, and including nothing that cannot at once be identified."—*Saturday Review.*

Astronomical Terms.

AN ASTRONOMICAL GLOSSARY; or, Dictionary of Terms used in Astronomy. With Tables of Data and Lists of Remarkable and Interesting Celestial Objects. By J. ELLARD GORE, F.R.A.S., Author of "The Visible Universe," &c. Small crown 8vo, 2s. 6d. cloth.

"A very useful little work for beginners in astronomy, and not to be despised by more advanced students."—*The Times.*

"A very handy book . . . the utility of which is much increased by its valuable tables of astronomical data."—*Athenæum.*

The Microscope.

THE MICROSCOPE: Its Construction and Management. Including Technique, Photo-micrography, and the Past and Future of the Microscope. By Dr. HENRI VAN HEURCK. Re-Edited and Augmented from the Fourth French Edition, and Translated by WYNNE E. BAXTER, F.G.S. 400 pages, with upwards of 250 Woodcuts, imp. 8vo, 18s., cloth.

"A translation of a well-known work, at once popular and comprehensive."—*Times.*

"The translation is as felicitous as it is accurate."—*Nature.*

The Microscope.

PHOTO-MICROGRAPHY. By Dr. H. VAN HEURCK. Extracted from the above Work. Royal 8vo, with Illustrations, 1s. sewed.

Astronomy.

ASTRONOMY. By the late Rev. ROBERT MAIN, M.A., F.R.S. Third Edition, Revised by WILLIAM THYNNE LYNN, B.A., F.R.A.S., formerly of the Royal Observatory, Greenwich. 12mo, 2s. cloth.

"A sound and simple treatise, very carefully edited, and a capital book for beginners."—*Knowledge.*

"Accurately brought down to the requirements of the present time by Mr. Lynn."—*Educational Times.*

Recent and Fossil Shells.

A MANUAL OF THE MOLLUSCA: Being a Treatise on Recent and Fossil Shells. By S. P. WOODWARD, A.L.S., F.G.S. With an Appendix on RECENT AND FOSSIL CONCHOLOGICAL DISCOVERIES by RALPH TATE, A.L.S., F.G.S. With 23 Plates and upwards of 300 Woodcuts. Reprint of Fourth Edition (1880). Crown 8vo, 7s. 6d. cloth.

"A most valuable storehouse of conchological and geological information."—*Science Gossip.*

Geology and Genesis.

THE TWIN RECORDS OF CREATION; or Geology and Genesis, their Perfect Harmony and Wonderful Concord. By G. W. V. LE VAUX. 8vo, 5s. cl.

'A valuable contribution to the evidences of Revelation, and disposes very conclusively of the arguments of those who would set God's Works against God's Word. No real difficulty is shirked, and no soil history is left unexposed."—*The Rock.*

Mechanics.

HANDBOOK OF MECHANICS. By Dr. LARDNER. Enlarged and re-written by BENJAMIN LOEWY, F.R.A.S. 378 Illusts. Post 8vo, 6s. cloth.

"The perspicuity of the original has been retained, and chapters which had become obsolete have been replaced by others of more modern character. The explanations throughout are studiously popular, and care has been taken to show the application of the various branches of physics to the industrial arts, and to the practical business of life."—*Mining Journal.*

Hydrostatics, &c.

HANDBOOK OF HYDROSTATICS AND PNEUMATICS. By Dr. LARDNER. New Edition, Revised and Enlarged by BENJAMIN LOEWY, F.R.A.S. With 236 Illustrations. Post 8vo, 5s. cloth.

"For those 'who desire to attain an accurate knowledge of physical science without the profound methods of mathematical investigation,' this work is well adapted."—*Chemical News.*

Heat.

HANDBOOK OF HEAT. By Dr. LARDNER. Edited and re-written by BENJAMIN LOEWY, F.R.A.S., &c. 117 Illustrations. Post 8vo, 6s. cloth.

"The style is always clear and precise, and conveys instruction without leaving any cloudiness or lurking doubts behind."—*Engineering.*

Optics.

HANDBOOK OF OPTICS. By Dr. LARDNER. New Edition. Edited by T. OLVER HARDING, B.A. Lond. With 298 Illustrations. Small 8vo, 448 pages, 5s. cloth.

"Written by one of the ablest English scientific writers, beautifully and elaborately illustrated." —*Mechanic's Magazine.*

Electricity, &c.

HANDBOOK OF ELECTRICITY, MAGNETISM, and Acoustics. By Dr. LARDNER. Edited by GEO. CAREY FOSTER, B.A., F.C.S. With 400 Illustrations. Small 8vo, 5s. cloth.

"The book could not have been entrusted to anyone better calculated to preserve the terse and lucid style of Lardner, while correcting his errors and bringing up his work to the present state of scientific knowledge."—*Popular Science Review.*

Astronomy.

HANDBOOK OF ASTRONOMY. By Dr. LARDNER. Fourth Edition. Revised and Edited by EDWIN DUNKIN, F.R.A.S., Royal Observatory, Greenwich. With 38 Plates and upwards of 100 Woodcuts. 8vo, 9s. 6d. cloth.

"Probably no other book contains the same amount of information in so compendious and well-arranged a form—certainly none at the price at which this is offered to the public."—*Athenæum.*

"We can do no other than pronounce this work a most valuable manual of astronomy, and we strongly recommend it to all who wish to acquire a general—but at the same time correct—acquaintance with this sublime science."—*Quarterly Journal of Science.*

Cyclopædia of Science.

MUSEUM OF SCIENCE AND ART. Edited by Dr. LARDNER. With upwards of 1,200 Engravings on Wood. In Six Double Volumes, £1 1s. in a new and elegant cloth binding; or handsomely bound in half-morocco, 31s. 6d.

"A cheap and interesting publication, alike informing and attractive. The papers combine subjects of importance and great scientific knowledge, considerable inductive powers, and a popular style of treatment."—*Spectator.*

Separate books formed from the above.

Common Things Explained. 5s.	**Steam and its Uses.** 2s. cloth.
The Microscope. 2s. cloth.	**Popular Astronomy.** 4s. 6d. cloth.
Popular Geology. 2s. 6d. cloth.	**The Bee and White Ants.** 2s. cloth.
Popular Physics. 2s. 6d. cloth.	**The Electric Telegraph.** 1s. 6d.

NATURAL PHILOSOPHY FOR SCHOOLS. By Dr. LARDNER. Fcap. 8vo, 3s. 6d.

"A very convenient class-book for junior students in private schools."—*British Quarterly Review.*

ANIMAL PHYSIOLOGY FOR SCHOOLS. By Dr. LARDNER. Fcap. 8vo, 3s. 6d.

"Clearly written, well arranged, and excellently illustrated."—*Gardener's Chronicle,*

THE ELECTRIC TELEGRAPH. By Dr. LARDNER. Revised by E. B. BRIGHT, F.R.A.S. Fcap. 8vo, 2s. 6d. cloth.

"One of the most readable books extant on the Electric Telegraph."—*English Mechanic.*

C

CHEMICAL MANUFACTURES, CHEMISTRY, etc.

Water.

WATER AND ITS PURIFICATION. A Handbook for the Use of Local Authorities, Sanitary Officers, and others interested in Water Supply. By SAMUEL RIDEAL, D.Sc., Lond., F.I.C. With numerous Illustrations and Tables. Crown 8vo, 7s. 6d. cloth. *[Just published.*

Chemistry for Engineers, etc.

ENGINEERING CHEMISTRY: A Practical Treatise for the Use of Analytical Chemists, Engineers, Iron Masters, Iron Founders, Students and others. Comprising Methods of Analysis and Valuation of the Principal Materials used in Engineering Work, with numerous Analyses, Examples and Suggestions. By H. JOSHUA PHILLIPS, F.I.C., F.C.S., Formerly Analytical and Consulting Chemist to the Great Eastern Railway. Second Edition, Revised and Enlarged. Crown 8vo, 400 pp., with Illustrations, 10s. 6d. cloth.

"In this work the author has rendered no small service to a numerous body of practical men. . . . The analytical methods may be pronounced most satisfactory, being as accurate as the despatch required of engineering chemists permits."—*Chemical News.*
"Those in search of a handy treatise on the subject of analytical chemistry as applied to the every-day requirements of workshop practice will find this volume of great assistance."—*Iron.*
"The book will be very useful to those who require a handy and concise *résumé* of approved methods of analysing and valuing metals, oils, fuels, &c. It is, in fact, a work for chemists, a guide to the routine of the engineering laboratory. . . . The book is full of good things. As a handbook of technical analysis, it is very welcome."—*Builder.*
"The analytical methods given are, as a whole, such as are likely to give rapid and trustworthy results in experienced hands. . . . There is much excellent descriptive matter in the work, the chapter on 'Oils and Lubrication' being specially noticeable in this respect."—*Engineer.*

Nitro-Explosives.

NITRO-EXPLOSIVES. A Practical Treatise concerning the Properties, Manufacture, and Analysis of Nitrated Substances, including the Fulminates, Smokeless Powders and Celluloid. By P. GERALD SANFORD, F.I.C. Consulting Chemist to the Cotton Powder Company, Limited, &c. With Illustrations. Crown 8vo, 9s. cloth. *[Just published,*

"Anyone having the requisite apparatus and materials could make nitro-glycerine or gun-cotton, to say nothing of other explosives, by the aid of the instructions contained in this volume. This is one of the very few text-books in which can be found just what is wanted. Mr. Sanford goes steadily through the whole list of explosives commonly used, he names any given explosive and tells us of what it is composed and how it is manufactured. The book is excellent throughout."—*The Engineer.*
"A good straightforward treatise on an important subject. Mr. Sanford has had unusual opportunities for making himself acquainted with the minutiæ of the manufacture of these explosives, and has set down his knowledge in a simple, easy, intelligible fashion that renders his book a valuable contribution to the literature of explosives."—*Mining Journal.*
"A thoroughly practical account of the manufacture, properties, &c., of nitro-explosives. It is an admirable work, which reflects the highest credit on the author."—*Colliery Guardian.*

Explosives.

A HANDBOOK ON MODERN EXPLOSIVES. A Practical Treatise on the Manufacture and Use of Dynamite, Gun-Cotton, Nitro-Glycerine and other Explosive Compounds, including Collodion-Cotton. With Chapters on Explosives in Practical Application. By M. EISSLER, Mining Engineer and Metallurgical Chemist, Author of "The Metallurgy of Gold," "The Metallurgy of Silver," &c. Second Edition, Enlarged. With 150 Illustrations. Crown 8vo, 12s. 6d. cloth. *[Just published.*

"Useful not only to the miner, but also to officers of both services to whom blasting and the use of explosives generally may at any time become a necessary auxiliary."—*Nature.*
"A veritable mine of information on the subject of explosives employed for military, mining and blasting purposes."—*Army and Navy Gazette.*

Explosives and Dangerous Goods.

DANGEROUS GOODS: Their Sources and Properties, Modes of Storage and Transport. With Notes and Comments on Accidents arising therefrom, together with the Government and Railway Classifications, Acts of Parliament, &c. A Guide for the Use of Government and Railway Officials, Steamship Owners, Insurance Companies and Manufacturers and Users of Explosives and Dangerous Goods. By H. JOSHUA PHILLIPS, F.I.C, F.C.S., Author of "Engineering, Chemistry," &c. Crown 8vo, 374 pages, 9s. cloth.

The Alkali Trade, Manufacture of Sulphuric Acid, &c.

A MANUAL OF THE ALKALI TRADE, including the Manufacture of Sulphuric Acid, Sulphate of Soda, and Bleaching Powder. By JOHN LOMAS, Alkali Manufacturer, Newcastle-upon-Tyne and London. With 232 Illustrations and Working Drawings, and containing 390 pages of Text. Second Edition, with Additions. Super-royal 8vo, £1 10s. cloth.

"This book is written by a manufacturer for manufacturers. The working details of the most approved forms of apparatus are given, and these are accompanied by no less than 232 wood engravings, all of which may be used for the purposes of construction. Every step in the manufacture is very fully described in this manual, and each improvement explained."—*Athenæum.*

"We find not merely a sound and luminous explanation of the chemical principles of the trade, but a notice of numerous matters which have a most important bearing on the successful conduct of alkali works, but which are generally overlooked by even experienced technological authors."—*Chemical Review.*

The Blowpipe.

THE BLOWPIPE IN CHEMISTRY, MINERALOGY, and Geology. Containing all known Methods of Anhydrous Analysis, many Working Examples, and Instructions for Making Apparatus. By Lieut.-Colonel W. A. ROSS, R.A., F.G.S. With 120 Illustrations. Second Edition, Enlarged. Crown 8vo, 5s. cloth.

"The student who goes conscientiously through the course of experimentation here laid down will gain a better insight into inorganic chemistry and mineralogy than if he had 'got up' any of the best text-books of the day, and passed any number of examinations in their contents."—*Chemical News.*

Commercial Chemical Analysis.

THE COMMERCIAL HANDBOOK of CHEMICAL ANALYSIS; or, Practical Instructions for the determination of the Intrinsic or Commercial Value of Substances used in Manufactures, in Trades, and in the Arts. By A. NORMANDY. New Edition by H. M. NOAD, Ph.D., F.R.S. Crown 8vo, 12s. 6d. cloth.

"We strongly recommend this book to our readers as a guide, alike indispensable to the housewife as to the pharmaceutical practitioner."—*Medical Times.*

Dye-Wares and Colours.

THE MANUAL OF COLOURS AND DYE-WARES: Their Properties, Applications, Valuations, Impurities, and Sophistications. For the use of Dyers, Printers, Drysalters, Brokers, &c. By J. W. SLATER. Second Edition, Revised and greatly Enlarged, crown 8vo, 7s. 6d. cloth.

"A complete encyclopædia of the *materia tinctoria.* The information given respecting each article is full and precise, and the methods of determining the value of articles such as these, so liable to sophistication, are given with clearness, and are practical as well as valuable."—*Chemist and Druggist.*

"There is no other work which covers precisely the same ground. To students preparing for examinations in dyeing and printing it will prove exceedingly useful."—*Chemical News.*

Modern Brewing and Malting.

A HANDYBOOK FOR BREWERS: Being a Practical Guide to the Art of Brewing and Malting. Embracing the Conclusions of Modern Research which bear upon the Practice of Brewing. By HERBERT EDWARDS WRIGHT. M.A. Second Edition, Enlarged. Crown 8vo, 530 pp., 12s. 6d. cloth.

"May be consulted with advantage by the student who is preparing himself for examinational tests, while the scientific brewer will find in it a *résumé* of all the most important discoveries of modern times. The work is written throughout in a clear and concise manner, and the author takes great care to discriminate between vague theories and well-established facts."—*Brewers' Journal.*

"We have great pleasure in recommending this handybook, and have no hesitation in saying that it is one of the best—if not the best—which has yet been written on the subject of beer-brewing in this country, it should have a place on the shelves of every brewer's library."—*Brewer's Guardian.*

"Although the requirements of the student are primarily considered, an acquaintance of half-an-hour's duration cannot fail to impress the practical brewer with the sense of having found a trustworthy guide and practical counsellor in brewery matters."—*Chemical Trade Journal.*

Analysis and Valuation of Fuels.

FUELS: SOLID, LIQUID, AND GASEOUS: Their Analysis and Valuation. For the Use of Chemists and Engineers. By H. J. PHILLIPS, F.C.S., Formerly Analytical and Consulting Chemist to the Great Eastern Railway. Third Edition, Revised and Enlarged. Crown 8vo, 2s. cloth.

"Ought to have its place in the laboratory of every metallurgical establishment, and wherever fuel is used on a large scale."—*Chemical News.*

Pigments.

THE ARTISTS' MANUAL OF PIGMENTS. Showing their Composition, Conditions of Permanency, Non-Permanency, and Adulterations ; Effects in Combination with Each Other and with Vehicles ; and the most Reliable Tests of Purity. By H. C. STANDAGE. Third Edition, crown 8vo, 2s. 6d. cloth.

" This work is indeed *multum-in-parvo,* and we can, with good conscience, recommend it to all who come in contact with pigments, whether as makers, dealers, or users."—*Chemical Review.*

Gauging.   Tables and Rules for Revenue Officers, Brewers, &c.

A POCKET-BOOK OF MENSURATION AND GAUGING : Containing Tables, Rules, and Memoranda for Revenue Officers, Brewers, Spirit Merchants, &c. By J. B. MANT (Inland Revenue). Second Edition, Revised. 18mo, 4s. leather.

" This handy and useful book is adapted to the requirements of the Inland Revenue Department, and will be a favourite book of reference. The range of subjects is comprehensive, and the arrangement simple and clear."—*Civilian.* " Should be in the hands of every practical brewer."—*Brewers' Journal.*

INDUSTRIAL ARTS, TRADES AND MANUFACTURES.

Cycle Manufacture.

MODERN CYCLES : A Practical Handbook on their Construction and Repair. By A. J. WALLIS-TAYLER, A.M.Inst.C.E. Author of " Refrigerating Machinery," &c. With upwards of 300 Illustrations. Crown 8vo, 10s. 6d. cloth.   [*Just published.*

Tea Growing and Manufacture.

TEA : Its History, Development, Culture, Botany, Manufacture, Chemistry, and the Laws especially affecting Labour. By DAVID CROLE. With Illustrative Plates and Plans. Medium 8vo, 16s. cloth.   [*Just published.*

Cotton Spinning.

COTTON MANUFACTURE : A Manual of Practical Instruction of the Processes of Opening, Carding, Combing, Drawing, Doubling and Spinning of Cotton, the Methods of Dyeing, &c. For the Use of Operatives, Overlookers, and Manufacturers. By JOHN LISTER, Technical Instructor, Pendleton. 8vo, 7s. 6d. cloth.

" This invaluable volume is a distinct advance in the literature of cotton manufacture."—*Machinery.*
" It is thoroughly reliable, fulfilling nearly all the requirements desired."—*Glasgow Herald.*

Flour Manufacture, Milling, etc.

FLOUR MANUFACTURE : A Treatise on Milling Science and Practice. By FRIEDRICH KICK, Imperial Regierungsrath, Professor of Mechanical Technology in the Imperial German Polytechnic Institute, Prague. Translated from the Second Enlarged and Revised Edition with Supplement. By H. H. P. POWLES, Assoc. Memb. Institution of Civil Engineers. Nearly 400 pp. Illustrated with 28 Folding Plates, and 167 Woodcuts. Roy. 8vo, 25s. cloth.

" This valuable work is, and will remain, the standard authority on the science of milling. . . : The miller who has read and digested this work will have laid the foundation, so to speak, of a successful career ; he will have acquired a number of general principles which he can proceed to apply. In this handsome volume we at last have the accepted text-book of modern milling in good, sound English, which has little, if any, trace of the German idiom."—*The Miller.*

" The appearance of this celebrated work in English is very opportune, and British millers will ; w are sure, not be slow in availing themselves of its pages."—*Millers' Gazette.*

Agglutinants.

CEMENTS, PASTES, GLUES AND GUMS: A Practical Guide to the Manufacture and Application of the various Agglutinants required in the Building, Metal-Working, Wood-Working, and Leather-Working Trades, and for Workshop, Laboratory or Office Use. With upwards of 900 Recipes and Formulæ. By H. C. STANDAGE, Chemist. Third Edition. Crown 8vo, 2s. cloth.

[*Just published.*

" We have pleasure in speaking favourably of this volume. So far as we have had experience, which is not inconsiderable, this manual is trustworthy."—*Athenæum.*

"As a revelation of what are considered trade secrets, this book will arouse an amount of curiosity among the large number of industries it touches."—*Daily Chronicle.*

Soap-making.

THE ART OF SOAP-MAKING: A Practical Handbook of the Manufacture of Hard and Soft Soaps, Toilet Soaps, &c. Including many New Processes, and a Chapter on the Recovery of Glycerine from Waste Leys. By ALEXANDER WATT. Fifth Edition, Revised, with an Appendix on Modern Candlemaking. Crown 8vo, 7s. 6d. cloth.

" The work will prove very useful, not merely to the technological student, but to the practical soap-boiler who wishes to understand the theory of his art."—*Chemical News.*

" A thoroughly practical treatise on an art which has almost no literature in our language. We congratulate the author on the success of his endeavour to fill a void in English technical literature."—*Nature.*

Paper Making.

PRACTICAL PAPER-MAKING: A Manual for Paper-Makers and Owners and Managers of Paper-Mills. With Tables, Calculations, &c. By G. CLAPPERTON, Paper-maker. With Illustrations of Fibres from Micro-photographs. Crown 8vo, 5s. cloth.

" The author caters for the requirements of responsible mill hands, apprentices, &c., whilst his manual will be found of great service to students of technology, as well as to veteran paper-makers and mill owners. The illustrations form an excellent feature."—*The World's Paper Trade Review.*

" We recommend everybody interested in the trade to get a copy of this thoroughly practical book."—*Paper Making.*

Paper Making.

THE ART OF PAPER-MAKING: A Practical Handbook of the Manufacture of Paper from Rags, Esparto, Straw, and other Fibrous Materials. Including the Manufacture of Pulp from Wood Fibre, with a Description of the Machinery and Appliances used. To which are added Details of Processes for Recovering Soda from Waste Liquors. By ALEXANDER WATT, Author of " The Art of Soap-Making." With Illustrations. Crown 8vo, 7s. 6d. cloth.

" It may be regarded as the standard work on the subject. The book is full of valuable information. The 'Art of Paper-making,' is in every respect a model of a text-book, either for a technical class, or for the private student."—*Paper and Printing Trades Journal.*

Leather Manufacture.

THE ART OF LEATHER MANUFACTURE: Being a Practical Handbook, in which the Operations of Tanning, Currying, and Leather Dressing are fully Described, and the Principles of Tanning Explained, and many Recent Processes Introduced; as also Methods for the Estimation of Tannin, and a Description of the Arts of Glue Boiling, Gut Dressing, &c. By ALEXANDER WATT, Author of " Soap-Making," &c. Fourth Edition. Crown 8vo, 9s. cloth.

" A sound, comprehensive treatise on tanning and its accessories. The book is an eminently valuable production, which redounds to the credit of both author and publishers."—*Chemical Review.*

Boot and Shoe Making.

THE ART OF BOOT AND SHOE-MAKING: A Practical Handbook, including Measurement, Last-Fitting, Cutting-Out, Closing and Making, with a Description of the most approved Machinery Employed. By JOHN B. LENO, late Editor of *St. Crispin,* and *The Boot and Shoe-Maker.* 12mo, 2s. cloth.

Dentistry Construction.

MECHANICAL DENTISTRY: A Practical Treatise on the Construction of the various kinds of Artificial Dentures. Comprising also Useful Formulæ, Tables, and Receipts for Gold Plate, Clasps, Solders, &c. &c. By C. HUNTER. Third Edition. With 100 Wood Engravings. Crown 8vo, 3s. 6d. cloth.

Wood Engraving.

WOOD ENGRAVING: A Practical and Easy Introduction to the Study of the Art. By W. N. BROWN. 12mo, 1s. 6d. cloth.

"The book is clear and complete, and will be useful to anyone wanting to understand the first elements of the beautiful art of wood engraving."—*Graphic.*

Horology.

A TREATISE ON MODERN HOROLOGY, in Theory and Practice. Translated from the French of CLAUDIUS SAUNIER, ex-Director of the School of Horology at Maçon, by JULIEN TRIPPLIN, F.R.A.S., Besancon Watch Manu'acturer, and EDWARD RIGG, M.A., Assayer in the Royal Mint. With Seventy-eight Woodcuts and Twenty-two Coloured Copper Plates. Second Edition. Super-royal 8vo, £2 2s. cloth; £2 10s. half-calf.

"There is no horological work in the English language at all to be compared to this production of M. Saunier's for clearness and completeness. It is alike good as a guide for the student and as a reference for the experienced horologist and skilled workman."—*Horological Journal.*

"The latest, the most complete, and the most reliable of those literary productions to which continental watchmakers are indebted for the mechanical superiority over their English brethren—in fact the Book of Books, is M. Saunier's 'Treatise.' "—*Watchmaker, Jeweller, and Silversmith.*

Watch Adjusting.

THE WATCH ADJUSTER'S MANUAL: A Practical Guide for the Watch and Chronometer Adjuster in Making, Springirg, Timing and Adjusting for Isochronism, Positions and Temperatures. By C. E. FRITTS. 370 pages, with Illustrations, 8vo, 16s. cloth.

Watchmaking.

THE WATCHMAKER'S HANDBOOK. Intended as a Workshop Companion for those engaged in Watchmaking and the Allied Mechanical Arts. Translated from the French of CLAUDIUS SAUNIER, and enlarged by JULIEN TRIPPLIN, F.R.A.S., and EDWARD RIGG, M.A., Assayer in the Royal Mint. Third Edition. Crown 8vo, 9s. cloth.

"Each part is truly a treatise in itself. The arrangement is good and the language is clear and concise. It is an admirable guide for the young watchmaker."—*Engineering.*

"It is impossible to speak too highly of its excellence. It fulfils every requirement in a handbook intended for the use of a workman. Should be found in every workshop."—*Watch and Clockmaker.*

Watches and Timekeepers.

A HISTORY OF WATCHES AND OTHER TIMEKEEPERS. By JAMES F. KENDAL, M.B.H. Inst. 1s. 6d. boards; or 2s. 6d. cloth, gilt.

"The best which has yet appeared on this subject in the English language."—*Industries.*

"Open the book where you may, there is interesting matter in it concerning the ingenious devices of the ancient or modern horologer."—*Saturday Review.*

Electrolysis of Gold, Silver, Copper, &c.

ELECTRO-DEPOSITION: A Practical Treatise on the Electrolysis of Gold, Silver, Copper, Nickel, and other Metals and Alloys. With descriptions of Voltaic Batteries, Magneto and Dynamo-Electric Machines, Thermopiles, and of the Materials and Processes used in every Department of the Art, and several Chapters on ELECTRO-METALLURGY. By ALEXANDER WATT, Author of "Electro-Metallurgy," &c. Third Edition, Revised. Crown 8vo, 9s., cloth.

"Eminently a book for the practical worker in electro-deposition. It contains practical descriptions of methods, processes and materials, as actually pursued and used in the workshop."—*Engineer.*

Electro-Metallurgy.

ELECTRO-METALLURGY: Practically Treated. By ALEXANDER WATT. Tenth Edition, including the most recent Processes. 12mo, 3s. 6d. cloth.

"From this book both amateur and artisan may learn everything necessary for the successful prosecution of electroplating."—*Iron.*

Working in Gold.

THE JEWELLER'S ASSISTANT in the Art of Working in Gold. A Practical Treatise for Masters and Workmen, Compiled from the Experience of Thirty Years' Workshop Practice. By GEORGE E. GEE, Author of "The Goldsmith's Handbook," &c. Crown 8vo, 7s. 6d. cloth.

"This manual of technical education is apparently destined to be a valuable auxiliary to a handcraft which is certainly capable of great improvement."—*The Times.*

Electroplating.

ELECTROPLATING : A Practical Handbook on the Deposition of Copper, Silver, Nickel, Gold, Aluminium, Brass, Platinum, &c. &c. By J. W. URQUHART, C.E. Third Edition, Revised. Crown 8vo, 5s. cloth.

"An excellent practical manual."—*Engineering.*
"An excellent work, giving the newest information."—*Horological Journal.*

Electrotyping.

ELECTROTYPING : The Reproduction and Multiplication of Printing Surfaces and Works of Art by the Electro-deposition of Metals. By J. W. URQUHART, C.E. Crown 8vo, 5s. cloth.

"The book is thoroughly practical ; the reader is, therefore, conducted through the leading laws of electricity, then through the metals used by electrotypers, the apparatus, and the depositing processes, up to the final preparation of the work."—*Art Journal.*

Goldsmiths' Work.

THE GOLDSMITH'S HANDBOOK. By GEORGE E. GEE, Jeweller, &c. Fourth Edition. 12mo, 3s. 6d. cloth.

"A good, sound educator, and will be generally accepted as an authority."—*Horological Journal.*

Silversmiths' Work.

THE SILVERSMITH'S HANDBOOK. By GEORGE E. GEE, Jeweller, &c. Third Edition, with numerous Illusts. 12mo, 3s. 6d. cloth.

"The chief merit of the work is its practical character. . . The workers in the trade will speedily discover its merits when they sit down to study it."—*English Mechanic.*

*** *The above two works together, strongly half-bound, price 7s.*

Sheet Metal Working.

SHEET METAL WORKER'S INSTRUCTOR : Comprising a Selection of Geometrical Problems and Practical Rules for Describing the Various Patterns Required by Zinc, Sheet-Iron, Copper and Tin-Plate Workers. By REUBEN HENRY WARN, Practical Tin-Plate Worker. New Edition, Revised and greatly Enlarged by JOSEPH G. HORNER, A.M.I.M.E. Crown 8vo, 254 pages, with 430 Illustrations. 7s. 6d. cloth. [*Just published.*

Bread and Biscuit Baking.

THE BREAD AND BISCUIT BAKER'S and Sugar-Boiler's Assistant. Including a large variety of Modern Recipes. With Remarks on the Art of Bread-making. By ROBERT WELLS. Third Edition. Cr. 8vo, 2s. cloth.

"A large number of wrinkles for the ordinary cook, as well as the baker."—*Saturday Review.*

Confectionery for Hotels and Restaurants.

THE PASTRYCOOK AND CONFECTIONER'S GUIDE. For Hotels, Restaurants, and the Trade in general, adapted also for Family Use. By R. WELLS, Author of "The Bread and Biscuit Baker." Crown 8vo, 2s. cloth.

"We cannot speak too highly of this really excellent work. In these days of keen competition our readers cannot do better than purchase this book."—*Baker's Times.*

Ornamental Confectionery.

ORNAMENTAL CONFECTIONERY : A Guide for Bakers, Confectioners and Pastrycooks ; including a Variety of Modern Recipes, and Remarks on Decorative and Coloured Work. With 129 Original Designs. By ROBERT WELLS. Crown 8vo, 5s. cloth.

"A valuable work, practical, and should be in the hands of every baker and confectioner. The illustrative designs are alone worth treble the amount charged for the whole work."—*Baker's Times.*

Flour Confectionery.

THE MODERN FLOUR CONFECTIONER, Wholesale and Retail. Containing a large Collection of Recipes for Cheap Cakes, Biscuits, &c. With Remarks on the Ingredients Used in their Manufacture. By ROBERT WELLS, Author of "The Bread and Biscuit Baker," &c. Crown 8vo, 2s. cloth.

"The work is of a decidedly practical character, and in every recipe regard is had to economical working."—*North British Daily Mail.*

Laundry Work.

LAUNDRY MANAGEMENT. A Handbook for Use in Private and Public Laundries. By the EDITOR of "The Laundry Journal." Third Edition. Crown 8vo, 2s. cloth.

"This book should certainly occupy an honoured place on the shelves of all housekeepers who wish to keep themselves *au courant* of the newest appliances and methods."—*The Queen.*

HANDYBOOKS FOR HANDICRAFTS.

BY PAUL N. HASLUCK,
Editor of "Work" (New Series), Author of "Lathe Work," "Milling Machines," &c.

Crown 8vo, 144 pages, cloth, price 1s. each.

☞ *These* HANDYBOOKS *have been written to supply information for* WORKMEN, STUDENTS, *and* AMATEURS *in the several Handicrafts, on the actual* PRACTICE *of the* WORKSHOP, *and are intended to convey in plain language* TECHNICAL KNOWLEDGE *of the several* CRAFTS. *In describing the processes employed, and the manipulation of material, workshop terms are used; workshop practice is fully explained; and the text is freely illustrated with drawings of modern tools, appliances, and processes.*

THE METAL TURNER'S HANDYBOOK. A Practical Manual for Workers at the Foot-Lathe. With over 100 Illustrations. Price 1s.

"The book will be of service alike to the amateur and the artisan turner. It displays thorough knowledge of the subje.t."—*Scotsman.*

THE WOOD TURNER'S HANDYBOOK. A Practical Manual for Workers at the Lathe. With over 100 Illustrations. Price 1s.

"We recommend the book to young turners and amateurs. A multitude of workmen have hitherto sought in vain for a manual of this special industry."—*Mechanical World.*

THE WATCH JOBBER'S HANDYBOOK. A Practical Manual on Cleaning, Repairing, and Adjusting. With upwards of 100 Illustrations. Price 1s.

"We strongly advise all young persons connected with the watch trade to acquire and study this inexpensive work."—*Clerkenwell Chronicle.*

THE PATTERN MAKER'S HANDYBOOK. A Practical Manual on the Construction of Patterns for Founders. With upwards of 100 Illustrations. 1s.

"A most valuable, if not indispensable, manual for the pattern maker."—*Knowledge.*

THE MECHANIC'S WORKSHOP HANDYBOOK. A Practical Manual on Mechanical Manipulation, embracing Information on various Handicraft Processes. With Useful Notes and Miscellaneous Memoranda. Comprising about 200 Subjects. Price 1s.

"A very clever and useful book, which should be found in every workshop; and it should certainly find a place in all technical schools."—*Saturday Review.*

THE MODEL ENGINEER'S HANDYBOOK. A Practical Manual on the Construction of Model Steam Engines. With upwards of 100 Illustrations. 1s.

"Mr. Hasluck has produced a very good little book."—*Builder.*

THE CLOCK JOBBER'S HANDYBOOK. A Practical Manual on Cleaning, Repairing, and Adjusting. With upwards of 100 Illustrations. Price 1s.

"It is of inestimable service to those commencing the trade."—*Coventry Standard.*

THE CABINET WORKER'S HANDYBOOK. A Practical Manual on the Tools, Materials, Appliances, and Processes employed in Cabinet Work. With upwards of 100 Illustrations. Price 1s.

"Mr. Hasluck's thoroughgoing little Handybook is amongst the most practical guides we have been for beginners in cabinet-work."—*Saturday Review.*

THE WOODWORKER'S HANDYBOOK OF MANUAL INSTRUCTION. Embracing Information on the Tools, Materials, Appliances and Processes Employed in Woodworking. With 104 Illustrations. Price 1s.

[Just published.

OPINIONS OF THE PRESS.

"Written by a man who knows, not only how work ought to be done, but how to do it, and how to convey his knowledge to others."—*Engineering.*

"Mr. Hasluck writes admirably, and gives complete instructions."—*Engineer.*

"Mr. Hasluck combines the experience of a practical teacher with the manipulative skill and scientific knowledge of processes of the trained mechanician, and the manuals are marvels of what can be produced at a popular price."—*Schoolmaster.*

"Helpful to workmen of all ages and degrees of experience."—*Daily Chronicle.*

"Practical, sensible, and remarkably cheap."—*Journal of Education.*

"Concise, clear, and practical."—*Saturday Review.*

COMMERCE, COUNTING-HOUSE WORK, TABLES, etc.

Commercial Education.

LESSONS IN COMMERCE. By Professor R. GAMBARO, of the
Royal High Commercial School at Genoa. Edited and Revised by JAMES GAULT, Professor of Commerce and Commercial Law in King's College, London. Second Edition, Revised. Crown 8vo, 3s. 6d. cloth.

"The publishers of this work have rendered considerable service to the cause of commercial education by the opportune production of this volume. . . . The work·is peculiarly acceptable to English readers and an admirable addition to existing class books. In a phrase, we think the work attains its object in furnishing a brief account of those laws and customs of British trade with which the commercial man interested therein should be familiar."—*Chamber of Commerce Journal.*

"An invaluable guide in the hands of those who are preparing for a commercial career, and, in fact the information it contains on matters of business should be impressed on every one."—*Counting House.*

Foreign Commercial Correspondence.

THE FOREIGN COMMERCIAL CORRESPONDENT: Being
Aids to Commercial Correspondence in Five Languages—English, French, German, Italian, and Spanish. By CONRAD E. BAKER. Second Edition. Cr. 8vo, 3s. 6d. cl.

"Whoever wishes to correspond in all the languages mentioned by Mr. Baker cannot do better than study this work, the materials of which are excellent and conveniently arranged. They consist not of entire specimen letters, but—what are far more useful—short passages, sentences, or phrases expressing the same general idea in various forms."—*Athenæum.*

"A careful examination h is convinced us that it is unusually complete, well arranged and reliable. The book is a thoroughly good one."—*Schoolmaster.*

Commercial French.

A NEW BOOK OF COMMERCIAL FRENCH: Grammar—
Vocabulary—Correspondence—Commercial Documents—Geography—Arithmetic—Lexicon. By P. CARROUÉ, Professor in the City High School J.—B. Say (Paris). Crown 8vo, 4s. 6d. cloth.

Accounts for Manufacturers.

FACTORY ACCOUNTS: Their Principles and Practice. A Hand-
book for Accountants and Manufacturers, with Appendices on the Nomenclature of Machine Details ; the Income Tax Acts ; the Rating of Factories ; Fire and Boiler Insurance ; the Factory and Workshop Acts, &c., including also a Glossary of Terms and a large number of Specimen Rulings. By EMILE GARCKE and J. M. FELLS. Fourth Edition, Revised and Enlarged. Demy 8vo, 250 pages. 6s. strongly bound.

"A very interesting description of the requirements of Factory Accounts. . . . The principle of assimilating the Factory Accounts to the general commercial books is one which we thoroughly agree with."—*Accountants' Journal.*

"Characterised by extreme thoroughness. There are few owners of factories who would not derive great benefit from the perusal of this most admirable work."—*Local Government Chronicle.*

Modern Metrical Units and Systems.

MODERN METROLOGY: A Manual of the Metrical Units and
Systems of the present Century. With an Appendix containing a proposed English System. By LOWIS D'A. JACKSON, A.-M. Inst. C.E., Author of "Aid to Survey Practice," &c. Large crown 8vo, 12s. 6d. cloth.

"We recommend the work to all interested in the practical reform of our weights and measures."
Nature.

The Metric System and the British Standards.

A SERIES OF METRIC TABLES, in which the British Standard
Measures and Weights are compared with those of the Metric System at present in Use on the Continent. By C. H. DOWLING, C.E. 8vo, 10s. 6d. strongly bound.

"Mr. Dowling's Tables are well put together as a ready reckoner for the conversion of one system into the other."—*Athenæum.*

Iron and Metal Trades' Calculator.

THE IRON AND METAL TRADES' COMPANION: For ex-
peditiously ascertaining the Value of any Goods bought or sold by Weight, from 1s. per cwt. to 112s. per cwt., and from one farthing per pound to one shilling per pound. By THOMAS DOWNIE. Strongly bound in leather, 396 pp., 9s.

"A most useful set of tables, nothing like them before existed."—*Building News.*

"Although specially adapted to the iron and metal trades, the tables will be found useful in every other business in which merchandise is bought and sold by weight."—*Railway News.*

Chadwick's Calculator for Numbers and Weights Combined:

THE NUMBER, WEIGHT, & FRACTIONAL CALCULATOR,

Containing upwards of 250,000 Separate Calculations, showing at a glance the value at 422 different rates, ranging from $\frac{1}{18}$th of a Penny to 20s. each, or per cwt., and £20 per ton, of any number of articles consecutively, from 1 to 470.—Any number of cwts., qrs., and lbs., from 1 cwt. to 470 cwts.—Any number of tons, cwts., qrs., and lbs., from 1 to 1,000 tons. By WILLIAM CHADWICK, Public Accountant. Third Edition, Revised and Improved. 8vo, 18s. strongly bound.

"It is as easy of reference for any answer or any number of answers as a dictionary. For making up accounts or estimates the book must prove invaluable to all who have any considerable quantity of calculations involving price and measure in any combination to do."—*Engineer.*

"The most perfect work of the kind yet prepared."—*Glasgow Herald.*

Harben's Comprehensive Weight Calculator.

THE WEIGHT CALCULATOR: Being a Series of Tables upon

a New and Comprehensive Plan, exhibiting at one Reference the exact Value of any Weight from 1 lb. to 15 tons, at 300 Progressive Rates, from 1d. to 168s. per cwt., and containing 186,000 Direct Answers, which, with their Combinations, consisting of a single addition (mostly to be performed at sight), will afford an aggregate of 10,266,000 Answers ; the whole being calculated and designed to ensure correctness and promote despatch. By HENRY HARBEN, Accountant. Fifth Edition, carefully corrected. Royal 8vo, strongly half-bound, £1 5s.

"A practical and useful work of reference for men of business generally."—*Ironmonger.* [*pendent.*

"Of priceless value to business men. It is a necessary book in all mercantile offices."—*Sheffield Inde-*

Harben's Comprehensive Discount Guide.

THE DISCOUNT GUIDE. Comprising several Series of Tables

for the use of Merchants, Manufacturers, Ironmongers, and others, by which may be ascertained the exact Profit arising from any mode of using Discounts, either in the Purchase or Sale of Goods, and the method of either Altering a Rate of Discount, or Advancing a Price, so as to produce, by one operation, a sum that will realise any required profit after allowing one or more Discounts : to which are added Tables of Profit or Advance from 1¼ to 90 per cent., Tables of Discount from 1¼ to 98¾ per cent., and Tables of Commission, &c., from ⅛ to 10 per cent. By HENRY HARBEN, Accountant. New Edition, Corrected. Demy 8vo, £1 5s. half-bound.

"A book such as this can only be appreciated by business men, to whom the saving of time means saving of money. The work must prove of great value to merchants, manufacturers, and general traders."—*British Trade Journal.*

A New Series of Calculators.

DIRECT CALCULATORS: A Series of Tables and Calculations

varied in arrangement to suit the needs of Particular Trades. By M. B. COTSWORTH. The Series comprises 13 distinct books, at prices ranging from 2s. 6d. to 10s. 6d. (Detailed prospectus on application).

New Wages Calculator.

TABLES OF WAGES at 54, 52, 50 and 48 Hours per Week.

Showing the Amounts of Wages from One-quarter-of-an-hour to Sixty-four hours in each case at Rates of Wages advancing by One Shilling from 4s. to 55s. per week. By THOS. GARBUTT, Accountant. Square crown 8vo, 6s. half-bound.

Iron Shipbuilders' and Merchants' Weight Tables.

IRON-PLATE WEIGHT TABLES: For Iron Shipbuilders,

Engineers, and Iron Merchants. Containing the Calculated Weights of upwards of 150,000 different sizes of Iron Plates from 1 foot by 6 in. by ¼ in. to 10 feet by 5 feet by 1 in. Worked out on the basis of 40 lbs. to the square foot of Iron of 1 inch in thickness. By H. BURLINSON and W. H. SIMPSON. 4to, 25s. half-bound.

Tables for Actuaries.

MATHEMATICAL TABLES (ACTUARIAL). Comprising

Commutation and Conversion Tables, Logarithms, Cologarithms, Antilogarithms and Reciprocals. By J. W. GORDON. Royal 8vo, mounted on canvas, in cloth case, 5s.
　　　　　　　　　　　　　　　　　　　　　　　　　　　　[*Just published.*

AGRICULTURE, FARMING, GARDENING, etc.

Dr. Fream's New Edition of "The Standard Treatise on Agriculture."

THE COMPLETE GRAZIER AND FARMER'S AND CATTLE BREEDER'S ASSISTANT: A Compendium of Husbandry. Originally Written by WILLIAM YOUATT. Thirteenth Edition, entirely Re-written, considerably Enlarged, and brought up to the Present Requirements of Agricultural Practice, by WILLIAM FREAM, LL.D., Steven Lecturer in the University of Edinburgh, Author of "The Elements of Agriculture," &c. Royal 8vo, 1,100 pp., with over 450 Illustrations. Price £1 11s. 6d. strongly and handsomely bound.

EXTRACT FROM PUBLISHERS' ADVERTISEMENT.

"A treatise that made its original appearance in the first decade of the century, and that enters upon its Thirteenth edition before the century has run its course, has undoubtedly established its position as a work of permanent value. . . . The phenomenal progress of the last dozen years in the Practice and Science of Farming has rendered it necessary, however, that the volume should be re-written, . . . and for this undertaking the Publishers were fortunate enough to secure the services of Dr. FREAM, whose high attainments in all matters pertaining to agriculture have been so emphatically recognised by the highest professional and official authorities. In carrying out his editorial duties, Dr. FREAM has been favoured with valuable contributions by Prof. J. WORTLEY AXE, Mr. E. BROWN, Dr. BERNARD DYER, Mr. W. J. MALDEN, Mr. R. H. REW, Prof. SHELDON, Mr. J. SINCLAIR, Mr. SANDERS SPENCER, and others.

"As regards the illustrations of the work, no pains have been spared to make them as representative and characteristic as possible, so as to be practically useful to the Farmer and Grazier."

SUMMARY OF CONTENTS.

BOOK I. ON THE VARIETIES, BREEDING, REARING, FATTENING AND MANAGEMENT OF CATTLE.

BOOK II. ON THE ECONOMY AND MANAGEMENT OF THE DAIRY.

BOOK III. ON THE BREEDING, REARING, AND MANAGEMENT OF HORSES.

BOOK IV. ON THE BREEDING, REARING, AND FATTENING OF SHEEP.

BOOK V. ON THE BREEDING, REARING, AND FATTENING OF SWINE.

BOOK VI. ON THE DISEASES OF LIVE STOCK.

BOOK VII. ON THE BREEDING, REARING, AND MANAGEMENT OF POULTRY.

BOOK VIII. ON FARM OFFICES AND IMPLEMENTS OF HUSBANDRY.

BOOK IX. ON THE CULTURE AND MANAGEMENT OF GRASS LANDS.

BOOK X. ON THE CULTIVATION AND APPLICATION OF GRASSES, PULSE AND ROOTS.

BOOK XI. ON MANURES AND THEIR APPLICATION TO GRASS LAND AND CROPS.

BOOK XII. MONTHLY CALENDARS OF FARMWORK.

*** OPINIONS OF THE PRESS ON THE NEW EDITION.

"Dr. Fream is to be congratulated on the successful attempt he has made to give us a work which will at once become the standard classic of the farm practice of the country. We believe that it will be found that it has no compeer among the many works at present in existence. . . . The illustrations are admirable, while the frontispiece, which represents the well-known bull, New Year's Gift, bred by the Queen, is a work of art."—*The Times.*

"The book must be recognised as occupying the proud position of the most exhaustive work of reference in the English language on the subject with which it deals."—*Athenæum.*

"The most comprehensive guide to modern farm practice that exists in the English language to-day. . . . The book is one that ought to be on every farm and in the library of every land owner."—*Mark Lane Express.*

"In point of exhaustiveness and accuracy the work will certainly hold a pre-eminent and unique position among books dealing with scientific agricultural practice. It is, in fact, an agricultural library of itself."—*North British Agriculturist.*

"A compendium of authoritative and well-ordered knowledge on every conceivable branch of the work of the live stock farmer; probably without an equal in this or any other country."—*Yorkshire Post.*

"The best and brightest guide to the practice of husbandry: one that has no superior—no equal we might truly say—among the agricultural literature now before the public. . . . In every section in which we have tested it, the work has been found thoroughly up to date."—*Bell's Weekly Messenger.*

British Farm Live Stock.

FARM LIVE STOCK OF GREAT BRITAIN. By ROBERT WALLACE, F.L.S., F.R.S.E., &c., Professor of Agriculture and Rural Economy in the University of Edinburgh. Third Edition, thoroughly Revised and considerably Enlarged. With over 120 Phototypes of Prize Stock. Demy 8vo, 384 pp., with 79 Plates and Maps. Price 12s. 6d., cloth.

"A really complete work on the history, breeds, and management of the farm stock of Great Britain, and one which is likely to find its way to the shelves of every country gentleman's library."—*The Times.*

"The latest edition of 'Farm Live Stock of Great Britain' is a production to be proud of, and its issue not the least of the services which its author has rendered to agricultural science."—*Scottish Farmer.*

"The book is very attractive, . . . and we can scarcely imagine the existence of a farmer who would not like to have a copy of this beautiful and useful work."—*Mark Lane Express.*

"A work which will long be regarded as a standard authority whenever a concise history and description of the breeds of live stock in the British Isles is required."—*Bell's Weekly Messenger.*

Dairy Farming.

BRITISH DAIRYING: A Handy Volume on the Work of the Dairy-Farm. For the Use of Technical Instruction Classes, Students in Agricultural Colleges and the Working Dairy-Farmer. By Prof. J. P. SHELDON, With Illusts. Second Edition, Revised. Crown 8vo, 2s. 6d. cloth. [*Just published.*

"Confidently recommended as a useful text-book on dairy farming. —*Agricultural Gazette.*
"Probably the best half-crown manual on dairy work that has yet been produced."—*North British Agriculturist.* "It is the soundest little work we have yet seen on the subject."—*The Times.*

Dairy Manual.

MILK, CHEESE, AND BUTTER: A Practical Handbook on their Properties and the Processes of their Production. Including a Chapter on Cream and the Methods of its Separation from Milk. By JOHN OLIVER, late Principal of the Western Dairy Institute, Berkeley. With Coloured Plates and 200 Illustrations. Crown 8vo, 7s. 6d. cloth.

"An exhaustive and masterly production. It may be cordially recommended to all students and practitioners of dairy science."—*N.B. Agriculturist.*
"We recommend this very comprehensive and carefully-written book to dairy-farmers and students of dairying. It is a distinct acquisition to the library of the agriculturist."—*Agricultural Gazette.*

Agricultural Facts and Figures.

NOTE-BOOK OF AGRICULTURAL FACTS AND FIGURES FOR FARMERS AND FARM STUDENTS. By PRIMROSE McCONNELL, B.Sc. Fifth Edition. Royal 32mo, roan, gilt edges, with band, 4s.

"Literally teems with information and we can cordially recommend it to all connected with agriculture."—*North British Agriculturist.*

Small Farming.

SYSTEMATIC SMALL FARMING; or, The Lessons of my Farm. Being an Introduction to Modern Farm Practice for Small Farmers. By R. SCOTT BURN, Author of "Outlines of Modern Farming," &c. Crown 8vo, 6s. cloth.

"This is the completest book of its class we have seen, and one which every amateur farmer will read with pleasure, and accept as a guide."—*Field.*

Modern Farming.

OUTLINES OF MODERN FARMING. By R. SCOTT BURN. Soils, Manures, and Crops—Farming and Farming Economy—Cattle, Sheep, and Horses—Management of Dairy, Pigs, and Poultry—Utilization of Town-Sewage, Irrigation, &c. Sixth Edition. In one vol., 1,250 pp., half-bound, profusely Illustrated, 12s.

Agricultural Engineering.

FARM ENGINEERING, THE COMPLETE TEXT-BOOK OF. Comprising Draining and Embanking; Irrigation and Water Supply; F Roads, Fences and Gates; Farm Buildings; Barn Implements and Machines; Fiel Implements and Machines; Agricultural Surveying, &c. By Professor JOHN SCOTT. In one vol., 1,150 pages, half-bound, with over 600 Illustrations, 12s.

"Written with great care, as well as with knowledge and ability. The author has done his work well; we have found him a very trustworthy guide wherever we have tested his statements. The volume will be of great value to agricultural students."—*Mark Lane Express.*

Agricultural Text-Book.

THE FIELDS OF GREAT BRITAIN: A Text-Book of Agricul ture. Adapted to the Syllabus of the Science and Art Department. For Elemen and Advanced Students. By HUGH CLEMENTS (Board of Trade). Secon Edition, Revised, with Additions. 18mo, 2s. 6d. cloth.

"It is a long time since we have seen a book which has pleased us more, or which contains such a vast and useful fund of knowledge."—*Educational Times.*

Tables for Farmers, &c.

TABLES AND MEMORANDA FOR FARMERS, GRAZIERS, Agricultural Students, Surveyors, Land Agents, Auctioneers, &c. With a Ne System of Farm Book-keeping. By SIDNEY FRANCIS. Third Edition, Revised 272 pp., waistcoat-pocket size, limp leather, 1s. 6d.

"Weighing less than 1 oz., and occupying no more space than a match box, it contains a mass o facts and calculations which has never before, in such handy form, been obtainable. Every opera tion on the farm is dealt with. The work may be taken as thoroughly accurate, the whole of th tables having been revised by Dr. Fream. We cordially recommend it."—*Bell's Weekly Messenger.*

Artificial Manures and Foods.

FERTILISERS AND FEEDING STUFFS : Their Properties and Uses. A Handbook for the Practical Farmer. By BERNARD DYER, D.Sc. (Lond.). With the Text of the Fertilisers and Feeding Stuffs Act of 1893, &c. Second Edition, Revised. Crown 8vo, 1s. cloth. [*Just published*.

The Management of Bees.

BEES FOR PLEASURE AND PROFIT : Guide to the Manipulation of Bees, the Production of Honey, and the General Management of the Apiary. By G. GORDON SAMSON. With numerous Illustrations. Crown 8vo, 1s. cloth.

Farm and Estate Book-keeping.

BOOK-KEEPING FOR FARMERS AND ESTATE OWNERS. A Practical Treatise, presenting, in Three Plans, a System adapted for all Classes of Farms. By JOHNSON M. WOODMAN, Chartered Accountant. Second Edition, Revised. Crown 8vo, 3s. 6d. cloth boards ; or, 2s. 6d. cloth limp.
" The volume is a capital study of a most important subject."—*Agricultural Gazette.*
" Farmers and land agents will find the book more than repay its cost and study."—*Building News.*

Farm Account Book.

WOODMAN'S YEARLY FARM ACCOUNT BOOK. Giving Weekly Labour Account and Diary, and showing the Income and Expenditure under each Department of Crops, Live Stock, Dairy, &c. &c. With Valuation, Profit and Loss Account, and Balance Sheet at the end of the Year. By JOHNSON M. WOODMAN, Chartered Accountant. Second Edition. Folio, half-bound, 7s. 6d. net.
" Contains every requisite form for keeping farm accounts readily and accurately."—*Agriculture.*

Early Fruits, Flowers and Vegetables.

THE FORCING-GARDEN ; or, How to Grow Early Fruits, Flowers, and Vegetables. With Plans and Estimates for Building Glasshouses, Pits and Frames. With Illustrations. By SAMUEL WOOD. Crown 8vo, 3s. 6d. cloth
" A good book, and fairly fills a place that was in some degree vacant. The book is written with great care, and contains a great deal of valuable teaching."—*Gardeners' Magazine.*

Good Gardening.

A PLAIN GUIDE TO GOOD GARDENING ; or, How to Grow Vegetables, Fruits, and Flowers. By S. WOOD. Fourth Edition, with considerable Additions, &c., and numerous Illustrations. Crown 8vo, 3s. 6d. cloth.
" A very good book, and one to be highly recommended as a practical guide. The practical directions are excellent."—*Athenæum.*

Gainful Gardening.

MULTUM-IN-PARVO GARDENING ; or, How to make One Acre of Land produce £620 a-year, by the Cultivation of Fruits and Vegetables ; also, How to Grow Flowers in Three Glass Houses, so as to realise £176 per annum clear Profit. By SAMUEL WOOD, Author of " Good Gardening," &c. Fifth and Cheaper Edition, revised, with Additions. Crown 8vo, 1s. sewed.
" We are bound to recommend it as not only suited to the case of the amateur and gentleman's gardener, but to the market grower."—*Gardeners' Magazine.*

Gardening for Ladies.

THE LADIES' MULTUM-IN-PARVO FLOWER GARDEN, and Amateur's Complete Guide. By S. WOOD. Crown 8vo, 3s. 6d. cloth.
" Full of shrewd hints and useful instructions, based on a lifetime of experience."—*Scotsman.*

Cultivation of the Potato.

POTATOES : How to Grow and Show Them. A Practical Guide to the Cultivation and General Treatment of the Potato. By J. PINK. Cr. 8vo, 2s.

Market Gardening.

MARKET AND KITCHEN GARDENING. By C. W. SHAW, late Editor of " Gardening Illustrated." 3s. 6d. cloth.
" The most valuable compendium of kitchen and market-garden work published."—*Farmer.*

AUCTIONEERING, VALUING, LAND SURVEYING, ESTATE AGENCY, etc.

Auctioneer's Assistant.

THE APPRAISER, AUCTIONEER, BROKER, House and Estate Agent and Valuer's Pocket Assistant, for the Valuation for Purchase, Sale, or Renewal of Leases, Annuities and Reversions, and of property generally ; with Prices for Inventories, &c. By JOHN WHEELER, Valuer, &c. Sixth Edition, Rewritten and greatly Extended by C. NORRIS, Surveyor, Valuer, &c. Royal 32mo, 5s. cloth.

"A neat and concise book of reference, containing an admirable and clearly-arranged list of prices for inventories, and a very practical guide to determine the value of furniture, &c."—*Standard.*

"Contains a large quantity of varied and useful information as to the valuation for purchase, sale, or renewal of leases, annuities and reversions, and of property generally, with prices for inventories, and a guide to determine the value of interior fittings and other effects."—*Builder.*

Auctioneering.

AUCTIONEERS : THEIR DUTIES AND LIABILITIES. A Manual of Instruction and Counsel for the Young Auctioneer. By ROBERT SQUIBBS, Auctioneer. Second Edition, Revised and partly Re-written. Demy 8vo, 12s. 6d. cl.

"The standard text-book on the topics of which it treats."—*Athenæum.*

"The work is one of general excellent character, and gives much information in a compendious and satisfactory form."—*Builder.*

"May be recommended as giving a great deal of information on the law relating to auctioneers, in a very readable form."—*Law Journal.*

"Auctioneers may be congratulated on having so pleasing a writer to minister to their special needs."—*Solicitors' Journal.*

Inwood's Estate Tables.

TABLES FOR THE PURCHASING OF ESTATES, Freehold, Copyhold, or Leasehold ; Annuities, Advowsons, &c., and for the Renewing of Leases held under Cathedral Churches, Colleges, or other Corporate bodies, for Terms of Years certain, and for Lives ; also for Valuing Reversionary Estates, Deferred Annuities, Next Presentations, &c. ; together with SMART's Five Tables of Compound Interest, and an Extension of the same to Lower and Intermediate Rates. By W. INWOOD. 24th Edition, with considerable Additions, and new and valuable Tables of Logarithms for the more Difficult Computations of the Interest of Money, Discount, Annuities, &c., by M. FÉDOR THOMAN, of the Société Crédit Mobilier of Paris. Crown 8vo, 8s. cloth.

"Those interested in the purchase and sale of estates, and in the adjustment of compensation cases, as well as in transactions in annuities, life insurances, &c., will find the present edition of eminent service."—*Engineering.*

"'Inwood's Tables' still maintain a most enviable reputation. The new issue has been enriched by large additional contributions by M. Fédor Thoman, whose carefully-arranged Tables cannot fail to be of the utmost utility."—*Mining Journal.*

Agricultural Valuer's Assistant.

THE AGRICULTURAL VALUER'S ASSISTANT. A Practical Handbook on the Valuation of Landed Estates ; including Rules and Data for Measuring and Estimating the Contents, Weights and Values of Agricultural Produce and Timber, and the Values of Feeding Stuffs, Manures, and Labour ; with Forms of Tenant-Right Valuations, Lists of Local Agricultural Customs, Scales of Compensation under the Agricultural Holdings Act, &c. &c. By TOM BRIGHT, Agricultural Surveyor. Second Edition, Enlarged. Crown 8vo, 5s. cloth.

"Full of tables and examples in connection with the valuation of tenant-right, estates, labour, contents and weights of timber, and farm produce of all kinds."—*Agricultural Gazette.*

"An eminently practical handbook, full of practical tables and data of undoubted interest and value to surveyors and auctioneers in preparing valuations of all kinds."—*Farmer.*

Plantations and Underwoods.

POLE PLANTATIONS AND UNDERWOODS: A Practical Handbook on Estimating the Cost of Forming, Renovating, Improving, and Grubbing Plantations and Underwoods, their Valuation for Purposes of Transfer, Rental, Sale or Assessment. By TOM BRIGHT. Crown 8vo, 3s. 6d. cloth.

"To valuers, foresters and agents it will be a welcome aid."—*North British Agriculturist.*

"Well calculated to assist the valuer in the discharge of his duties, and of undoubted interest and use both to surveyors and auctioneers in preparing valuations of all kinds."—*Kent Herald.*

Hudson's Land Valuer's Pocket-Book.

THE LAND VALUER'S BEST ASSISTANT : Being Tables on a very much improved Plan, for Calculating the Value of Estates. With Tables for reducing Scotch, Irish, and Provincial Customary Acres to Statute Measure, &c. By R. HUDSON, C.E. New Edition. Royal 32mo, leather, elastic band, 4s.
"Of incalculable value to the country gentleman and professional man."—*Farmers' Journal.*

Ewart's Land Improver's Pocket-Book.

THE LAND IMPROVER'S POCKET-BOOK of Formulæ, Tables, and Memoranda required in any Computation relating to the Permanent Improvement of Landed Property. By JOHN EWART, Surveyor. Second Edition, Revised. Royal 32mo, oblong, 4s. leather.
"A compendious and handy little volume."—*Spectator.*

Complete Agricultural Surveyor's Pocket-Book.

THE LAND VALUER'S and Land Improver's Complete Pocket-Book. Being the above Two Works bound together. 7s. 6d. leather.

House Property.

HANDBOOK OF HOUSE PROPERTY : A Popular and Practical Guide to the Purchase, Mortgage, Tenancy, and Compulsory Sale of Houses and Land, including the Law of Dilapidations and Fixtures: with Examples of all kinds of Valuations, Useful Information on Building and Suggestive Elucidations of Fine Art. By E. L. TARBUCK, Architect and Surveyor. Fifth Edition, Enlarged. 12mo, 5s. cloth.

LAW AND MISCELLANEOUS.

Journalism.

MODERN JOURNALISM : A Handbook of Instruction and Counsel for the Young Journalist. By JOHN B. MACKIE, Fellow of the Institute of Journalists. Crown 8vo, 2s. cloth.
"This invaluable guide to journalism is a work which all aspirants to a journalistic career will read with advantage."—*Journalist.*

Private Bill Legislation and Provisional Orders.

HANDBOOK FOR THE USE OF SOLICITORS and Engineers Engaged in Promoting Private Acts of Parliament and Provisional Orders, for the authorization of Railways, Tramways, Gas and Water Works, &c. By L. LIVINGSTON MACASSEY, of the Middle Temple, Barrister-at-Law, M.Inst.C.E. 8vo, 25s. cloth.

Law of Patents.

PATENTS FOR INVENTIONS, and How to Procure Them. Compiled for the Use of Inventors, Patentees and others. By G. G. M. HARDINGHAM, Assoc.Mem.Inst.C.E., &c. Demy 8vo, 1s. 6d. cloth.

Labour Disputes.

CONCILIATION AND ARBITRATION in Labour Disputes : A Historical Sketch and Brief Statement of the Present Position of the Question at Home and Abroad. By J. S. JEANS, Author of "England's Supremacy," &c. Crown 8vo, 200 pp., 2s. 6d. cloth.

Pocket-Book for Sanitary Officials.

THE HEALTH OFFICER'S POCKET-BOOK : A Guide to Sanitary Practice and Law. For Medical Officers of Health, Sanitary Inspectors, Members of Sanitary Authorities, &c. By EDWARD F. WILLOUGHBY, M.D. (Lond.), &c. Fcap. 8vo, 7s. 6d., cloth.
"A mine of condensed information of a pertinent and useful kind on the various subjects of which it treats. The matter seems to have been carefully compiled and arranged for facility of reference, and it is well illustrated by diagrams and woodcuts. The different subjects are succinctly but fully and scientifically dealt with."—*The Lancet.*
"Ought to be welcome to those for whose use it is designed, since it practically boils down a reference library into a pocket volume. . . . It combines, with an uncommon degree of efficiency, the qualities of accuracy, conciseness and comprehensiveness."—*Scotsman.*

A Complete Epitome of the Laws of this Country.

EVERY MAN'S OWN LAWYER: A Handy-Book of the Principles of Law and Equity. With a Concise Dictionary of Legal Terms. By A BARRISTER. Thirty-fourth Edition, carefully Revised, and including New Acts of Parliament of 1896. Comprising the *Locomotives on Highways Act,* 1896, *amending the law to meet the case of the New Motor Cars or Horseless Carriages ; the Finance Act,* 1896 ; *the Agricultural Rates Act,* 1896 ; *the Conciliation Act,* 1896 ; *the Truck Act,* 1896 ; *the Light Railways Act,* 1896 ; *the Judicial Trustees Act,* 1896 ; *the Vexatious Actions Act,* 1896 ; *the Wild Birds Protection Act,* 1896 ; *the London Cab Act,* 1896, &c. *Judicial Decisions during the year have also been duly noted.* Crown 8vo, 750 pp., price 6*s.* 8*d.* (saved at every consultation !), strongly bound in cloth. [*Just published.*

*** *The Book will be found to comprise (amongst other matter)—*

THE RIGHTS AND WRONGS OF INDIVIDUALS—LANDLORD AND TENANT— VENDORS AND PURCHASERS — LEASES AND MORTGAGES — PRINCIPAL AND AGENT — PARTNERSHIP AND COMPANIES — MASTERS, SERVANTS AND WORK- MEN—CONTRACTS AND AGREEMENTS—BORROWERS, LENDERS AND SURETIES— SALE AND PURCHASE OF GOODS — CHEQUES, BILLS AND NOTES — BILLS OF SALE — BANKRUPTCY — RAILWAY AND SHIPPING LAW — LIFE, FIRE, AND MARINE INSURANCE—ACCIDENT AND FIDELITY INSURANCE—CRIMINAL LAW— PARLIAMENTARY ELECTIONS — COUNTY COUNCILS — DISTRICT COUNCILS— PARISH COUNCILS—MUNICIPAL CORPORATIONS—LIBEL AND SLANDER—PUBLIC HEALTH AND NUISANCES — COPYRIGHT, PATENTS, TRADE MARKS — HUSBAND AND WIFE — DIVORCE — INFANCY — CUSTODY OF CHILDREN — TRUSTEES AND EXECUTORS — CLERGY, CHURCHWARDENS, ETC. — GAME LAWS AND SPORTING— INNKEEPERS — HORSES AND DOGS — TAXES AND DEATH DUTIES — FORMS OF AGREEMENTS, WILLS, CODICILS, NOTICES, ETC.

☞ *The object of this work is to enable those who consult it to help themselves to the law ; and thereby to dispense, as far as possible, with professional assistance and advice. There are many wrongs and grievances which persons submit to from time to time through not knowing how or where to apply for redress ; and many persons have as great a dread of a lawyer's office as of a lion's den. With this book at hand it is believed that many a* SIX-AND-EIGHTPENCE *may be saved ; many a wrong redressed ; many a right reclaimed ; many a law suit avoided ; and many an evil abated. The work has established itself as the standard legal adviser of all classes, and has also made a reputation for itself as a useful book of reference for lawyers residing at a distance from law libraries, who are glad to have at hand a work embodying recent decisions and enactments.*

*** OPINIONS OF THE PRESS.

" It is a complete code of English Law written in plain language, which all can understand. . . . Should be in the hands of every business man, and all who wish to abolish lawyers' bills."—*Weekly Times.*

" A useful and concise epitome of the law, compiled with considerable care."—*Law Magazine.*

" A complete digest of the most useful facts which constitute English law."—*Globe.*

" This excellent handbook. . . . Admirably done, admirably arranged, and admirably cheap."— *Leeds Mercury.*

" A concise, cheap, and complete epitome of the English law. So plainly written that he who runs may read, and he who reads may understand."—*Figaro.*

" A dictionary of legal facts well put together. The book is a very useful one."—*Spectator.*

Legal Guide for Pawnbrokers.

THE PAWNBROKERS', FACTORS', AND MERCHANTS' GUIDE TO THE LAW OF LOANS AND PLEDGES. With the Statutes and a Digest of Cases. By H. C. FOLKARD, Barrister-at-Law. 3*s.* 6*d.* cloth.

The Law of Contracts.

LABOUR CONTRACTS: A Popular Handbook on the Law of Contracts for Works and Services. By DAVID GIBBONS. Fourth Edition, with Appendix of Statutes by T. F. UTTLEY, Solicitor. Fcap. 8vo, 3*s.* 6*d.* cloth.

The Factory Acts.

SUMMARY OF THE FACTORY AND WORKSHOP ACTS (1878-1891). For the Use of Manufacturers and Managers. By EMILE GARCKE and J. M. FELLS. (Reprinted from " FACTORY ACCOUNTS.") Cr. 8vo, 6*d.* sewed.

WEALE'S SERIES

OF

SCIENTIFIC & TECHNICAL

WORKS.

"It is not too much to say that no books have ever proved more popular with or more useful to young engineers and others than the excellent treatises comprised in WEALE'S SERIES."—**Engineer.**

A New Classified List.

CROSBY LOCKWOOD AND SON,

7, STATIONERS' HALL COURT, LONDON, E.C.

1897.

CIVIL ENGINEERING & SURVEYING.

Civil Engineering.

By HENRY LAW, M. Inst. C.E. Including a Treatise on HYDRAULIC ENGINEERING by G. R. BURNELL, M.I.C.E. Seventh Edition, revised, WITH LARGE ADDITIONS ON RECENT PRACTICE by D. KINNEAR CLARK, M. Inst. C.E. **6/6**, cloth boards **7/6**.
" An admirable volume, which we warmly recommend to young engineers."—*Builder.*

Pioneer Engineering.

A Treatise on the Engineering Operations connected with the Settlement of Waste Lands in New Countries. By E. DOBSON, A.I.C.E. Second Edition . . **4/6**
" Mr. Dobson is familiar with the difficulties which have to be overcome in this class of work, and much of his advice will be valuable to young engineers proceeding to our colonies."—*Engineering.*

Iron Bridges of Moderate Span:

Their Construction and Erection. By H. W. PENDRED. With 40 illustrations **2/0**
" Students and engineers should obtain this book for constant and practical use."—*Colliery Guardian.*

A Treatise on the Application of Iron to the Construction of Bridges, Roofs, & other Works.

By FRANCIS CAMPIN, C.E. Fourth Edition **2/6**
" For numbers of young engineers the book is just the cheap, handy, first guide they want."—*Middlesbrough Weekly News.* " Remarkably accurate and well written."—*Artisan.*

Constructional Iron and Steel Work,

As applied to Public, Private, and Domestic Buildings. By FRANCIS CAMPIN C.E. **3/6**
" This practical book may be counted a most valuable work."—*British Architect.*

Tubular and other Iron Girder Bridges,

Describing the Britannia and Conway Tubular Bridges. With a Sketch of Iron Bridges, &c. By G. DRYSDALE DEMPSEY, C.E. Fourth Edition . . **2/0**

Materials and Construction.

A Theoretical and Practical Treatise on the Strains, Designing, and Erection of Works of Construction. By FRANCIS CAMPIN, C.E. Second Edition . **3/0**
" No better exposition of the practical application of the principles of construction has yet been published to our knowledge in such a cheap comprehensive form."—*Building News.*

Sanitary Work in Small Towns and Villages.

By CHARLES SLAGG, Assoc. M. Inst. C.E. Second Edition, Enlarged . . **3/0**
" This is a very useful book. There is a great deal of work required to be done in the smaller towns and villages, and this little volume will help those who are willing to do it."—*Builder.*

Construction of Roads and Streets.

In Two Parts: I. THE ART OF CONSTRUCTING COMMON ROADS, by H. LAW, C.E., Revised by D. KINNEAR CLARK, C.E.; II. RECENT PRACTICE: Including Pavements of Stone, Wood, and Asphalte. By D. K. CLARK, C.E. . . **4/6**
" A book which every borough surveyor and engineer must possess, and which will be of considerable service to architects, builders, and property owners generally."—*Building News.*

Construction of Gas Works,

And the Manufacture and Distribution of Coal Gas. By S. HUGHES, C.E. Re-written by WILLIAM RICHARDS. C.E. Eighth Edition, with important Additions . **5/6**
" Will be of infinite service alike to manufacturers, distributors, and consumers."—*Foreman Engineer.*

Water Works, for the Supply of Cities and Towns.

With a Description of the Principal Geological Formations of England as influencing Supplies of Water. By SAMUEL HUGHES **4/0**
" Everyone who is debating how his village, town, or city shall be plentifully supplied with pure water should read this book.—*Newcastle Courant.*

Power of Water.

As applied to drive Flour Mills, and to give motion to Turbines and other Hydrostatic Engines. By JOSEPH GLYNN, F.R.S., &c. New Edition. Illustrated . . **2/0**

Wells and Well-Sinking.

By J. G. SWINDELL, A.R.I.B.A., and G. R. BURNELL, C.E. Revised Edition **2/0**
" Solid practical information, written in a concise and lucid style. The work can be recommended as a text-book for all surveyors, architects &c."—*Iron and Coal Trades Review.*

Drainage of Lands, Towns, and Buildings

By G. D. DEMPSEY, C.E. Revised, with large Additions on Recent Practice in Drainage Engineering, by D. KINNEAR CLARK, M.I.C.E. Third Edition . **4/6**

Embanking Lands from the Sea.

With Examples and Particulars of actual Embankments, &c. By JOHN WIGGINS, F.G.S. **2/0**

Blasting and Quarrying of Stone,

For Building and other Purposes. With Remarks on the Blowing up of Bridges. By Gen. Sir J. BURGOYNE, K.C.B. **1/6**

Foundations and Concrete Works.

With Practical Remarks on Footings, Planking, Sand, Concrete. Béton, Pile-driving, Caissons, and Cofferdams. By E. DOBSON, M.R.I.B.A. Seventh Edition . **1/6**

Pneumatics,

Including Acoustics and the Phenomena of Wind Currents, for the use of Beginners. By CHARLES TOMLINSON, F.R.S. Fourth Edition, enlarged. Illustrated . **1/6**

Land and Engineering Surveying.

For Students and Practical Use. By T. BAKER, C.E. Fifteenth Edition, revised and corrected by J. R. YOUNG, formerly Professor of Mathematics, Belfast College. Illustrated with Plates and Diagrams **2/0**

Mensuration and Measuring,

With the Mensuration and Levelling of Land for the purposes of Modern Engineering. By. T. BAKER, C.E. New Edition by E. NUGENT, C.E. **1/6**

MINING AND METALLURGY.

Mineralogy,

Rudiments of. By A. RAMSAY, F.G.S. Third Edition. Woodcuts and Plates. **3/6**
"The author throughout has displayed an intimate knowledge of his subject, and great facility in imparting that knowledge to others. The book is of great utility."—*Mining Journal.*

Coal and Coal Mining,

A Rudimentary Treatise on. By the late Sir WARINGTON W. SMYTH, M.A., F.R.S., &c., Chief Inspector of the Mines of the Crown. 7th Edn., revised and enlarged **3/6**
"Every portion of the volume appears to have been prepared with much care; and as an outline is given of every known coal-field in this and other countries, as well as of the two principal methods of working, the book will doubtless interest a very large number of readers."—*Mining Journal.*

Metallurgy of Iron.

Containing History of Iron Manufacture, Methods of Assay, and Analyses of Iron Ores, Processes of Manufacture of Iron and Steel, &c. By H. BAUERMAN, F.G.S., A.R.S.M. With numerous Illustrations. Sixth Edition, revised and enlarged . . **5/0**
"Carefully written, it has the merit of brevity and conciseness, as to less important points; while all material matters are very fully and thoroughly entered into."—*Standard.*

Mineral Surveyor & Valuer's Complete Guide.

Comprising a Treatise on Improved Mining Surveying and the Valuation of Mining Properties, with New Traverse Tables. By W. LINTERN, C.E., Third Edition, with an Appendix on Magnetic and Angular Surveying, with Records of the Peculiarities of Needle Disturbances. With Four Plates of Diagrams, Plans, &c. . . . **3/6**
"Contains much valuable information, and is thoroughly trustworthy."—*Iron & Coal Trades Review.*

Slate and Slate Quarrying,

Scientific, Practical, and Commercial. By D. C. DAVIES, F.G.S., Mining Engineer, &c. With numerous Illustrations and Folding Plates. Third Edition . . . **3/0**
"One of the best and best-balanced treatises on a special subject that we have met with."—*Engineer.*

A First Book of Mining and Quarrying,

With the Sciences connected therewith, for Primary Schools and Self-Instruction. By J. H. COLLINS, F.G.S. Second Edition, with Additions **1/6**
"For those concerned in schools in the mining districts, this work is the very thing that should be in the hands of their schoolmasters."—*Iron.*

Subterraneous Surveying.

By THOMAS FENWICK. Also the Method of Conducting Subterraneous Surveys without the use of the Magnetic Needle, &c. By T. BAKER, C.E. . . . **2/6**

Mining Tools,

Manual of. By WILLIAM MORGANS, Lecturer on Practical Mining at the Bristol School of Mines **2/6**

Mining Tools, Atlas

Of Engravings to Illustrate the above, containing 235 Illustrations of Mining Tools, drawn to Scale. 4to. **4/6**
"Students, Overmen, Captains, Managers, and Viewers may gain practical knowledge and useful hints by the study of Mr. Morgans' Manual."—*Colliery Guardian.*

Physical Geology,

Partly based on Major-General PORTLOCK's " Rudiments of Geology." By RALPH TATE, A.L.S., &c. Woodcuts **2/0**

Historical Geology,

Partly based on Major-General PORTLOCK's " Rudiments." By RALPH TATE . **2/6**

Geology,

PHYSICAL and HISTORICAL. Consisting of " Physical Geology," which sets forth the Leading Principles of the Science ; and " Historical Geology," which treats of the Mineral and Organic Conditions of the Earth at each successive epoch. By RALPH TATE, F.G.S. With 250 Illustrations **4/0**
" The fulness of the matter has elevated the book into a manual. Its information is exhaustive and well-arranged, so that any subject may be opened upon at once."—*School Board Chronicle.*

MECHANICAL ENGINEERING.

Workman's Manual of Engineering Drawing.

By JOHN MAXTON, Instructor in Engineering Drawing, Royal Naval College, Greenwich. Seventh Edition. 300 Plates and Diagrams **3/6**
" A copy of it should be kept for reference in every drawing office."—*Engineering.*

Fuels : Solid, Liquid, and Gaseous.

Their Analysis and Valuation. For the use of Chemists and Engineers. By H. J. PHILLIPS, F.C.S., formerly Analytical and Consulting Chemist to the Great Eastern Railway. Second Edition, revised **2/0**
" Ought to have its place in the laboratory of every metallurgical establishment, and wherever fuel is used on a large scale."—*Chemical News.*

Fuel, Its Combustion and Economy.

Consisting of an Abridgment of " A Treatise on the Combustion of Coal and the Prevention of Smoke." By C. W. WILLIAMS, A.I.C.E. With extensive Additions by D. KINNEAR CLARK, M. Inst. C.E. Third Edition, corrected . . . **3/6**
" Students should buy the book and read it, as one of the most complete and satisfactory treatises on the combustion and economy of fuel to be had.'—*Engineer.*

Boilermaker's Assistant

In Drawing, Templating, and Calculating Boiler Work, &c. By J. COURTNEY, Practical Boilermaker. Edited by D. K. CLARK, C.E. Third Edition, revised **2/0**
" With very great care we have gone through the ' Boilermaker's Assistant,' and have to say that it has our unqualified approval. Scarcely a point has been omitted."—*Foreman Engineer.*

Boilermaker's Ready Reckoner,

With Examples of Practical Geometry and Templating for the Use of Platers, Smiths, and Riveters. By JOHN COURTNEY. Edited by D. K. CLARK, M.I.C.E. . **4/0**
..* The last two Works in One Vol., half-bound, entitled " THE BOILERMAKER'S READY RECKONER AND ASSISTANT." By J. COURTNEY and D. K. CLARK. *Price* **7/0**
" A most useful work. No workman or apprentice should be without it."—*Iron Trade Circular.*

Steam Boilers.

Their Construction and Management. By R. ARMSTRONG, C.E. Illustrated . **1/6**
" A mass of information suitable for beginners."—*Design and Work.*

Steam and Machinery Management.

A Guide to the Arrangement and Economical Management of Machinery, with Hints on Construction and Selection. By M. POWIS BALE, M. Inst. M.E. . . . **2/6**
" Gives the results of wide experience."—*Lloyd's Newspaper.*

Steam and the Steam Engine.

Stationary and Portable. Being an Extension of the Treatise on the Steam Engine of Mr. J. SEWELL. By D. K. CLARK, C.E. Third Edition **3/6**
" Every essential part of the subject is treated of competently, and in a popular style."—*Iron.*

The Steam Engine,

A Treatise on the Mathematical Theory of, with Rules and Examples for Practical Men. By T. BAKER, C.E. **1/6**
" Teems with scientific information with reference to the steam-engine."—*Design and Work.*

The Steam Engine.

For the use of Beginners. By Dr. LARDNER. Illustrated **1/6**

Locomotive Engines.

A Rudimentary Treatise on. By G. D. DEMPSEY, C.E. With large Additions treating of the Modern Locomotive, by D. K. CLARK, M. Inst. C.E. With Illustrations **3/0**
" A model of what an elementary technical book should be."—*Academy.*

Locomotive Engine-Driving.

A Practical Manual for Engineers in Charge of Locomotive Engines. By MICHAEL REYNOLDS, M.S.E. Eighth Edition. 3/6 limp ; cloth boards . . 4/6

"We can confidently recommend the book, not only to the practical driver, but to everyone who takes an interest in the performance of locomotive engines."—*The Engineer.*

Stationary Engine Driving.

A Practical Manual for Engineers in Charge of Stationary Engines. By MICHAEL REYNOLDS, M.S.E. Fourth Edition. 3/6 limp : cloth boards . . 4/6

"The author is thoroughly acquainted with his subjects, and has produced a manual which is an exceedingly useful one for the class for whom it is specially intended."—*Engineering.*

Smithy and Forge.

Including the Farrier's Art and Coach Smithing. By W. J. E. CRANE. . 2/6

"The first modern English book on the subject. Great pains have been bestowed by the author upon the book ; shoeing-smiths will find it both useful and interesting."—*Builder.*

Modern Workshop Practice,

As applied to Marine, Land, and Locomotive Engines, Floating Docks, Dredging Machines, Bridges, Ship-Building, &c. By J. G. WINTON. 4th Edn., Illustrated 3/6

"Whether for the apprentice determined to master his profession, or for the artisan bent upon raising himself to a higher position, this clearly-written and practical treatise will be a great help."—*Scotsman.*

Mechanical Engineering.

Comprising Metallurgy, Moulding, Casting, Forging, Tools, Workshop Machinery, Mechanical Manipulation, Manufacture of the Steam Engine, &c. By FRANCIS CAMPIN, C.E. Third Edition, Re-written and Enlarged . . [*Just published.* 2/6

"A sound and serviceable text-book, quite up to date."—*Building News.*

Details of Machinery.

Comprising Instructions for the Execution of various Works in Iron in the Fitting-shop, Foundry, and Boiler-Yard. By FRANCIS CAMPIN, C.E. . . . 3/0

"A sound and practical handbook for all engaged in the engineering trades."—*Building World.*

Elementary Engineering :

A Manual for Young Marine Engineers and Apprentices. In the Form of Questions and Answers on Metals, Alloys, Strength of Materials, &c. By J. S. BREWER. Second Edition 2/0

"A useful introduction to the more elaborate text-books."—*Scotsman.*

Power in Motion :

Horse-power Motion, Toothed-Wheel Gearing, Long and Short Driving Bands, Angular Forces, &c. By JAMES ARMOUR, C.E. With 73 Diagrams. Third Edition . 2/0

"The value of the knowledge imparted cannot well be over-estimated."—*Newcastle Weekly Chron.*

Iron and Heat.

Exhibiting the Principles concerned in the Construction of Iron Beams, Pillars and Girders. By J. ARMOUR, C.E. 2/6

"A very useful and thoroughly practical little volume."—*Mining Journal.*

Practical Mechanism,

And Machine Tools. By T. BAKER, C.E. With Remarks on Tools and Machinery by J. NASMYTH, C.E. 2/6

Mechanics.

Being a concise Exposition of the General Principles of Mechanical Science, and their Applications. By CHARLES TOMLINSON, F.R.S. 1/6

Cranes,

The Construction of, and other Machinery for Raising Heavy Bodies for the Erection of Buildings, &c. By JOSEPH GLYNN, F.R.S. 1/6

NAVIGATION, SHIPBUILDING, ETC.

Sailor's Sea Book :

A Rudimentary Treatise on Navigation. By JAMES GREENWOOD, B.A. With numerous Woodcuts and Coloured Plates. New and Enlarged Edition. By W. H. ROSSER 2/6

"Is perhaps the best and simplest epitome of navigation ever compiled."—*Field.*

Practical Navigation.

Consisting of the SAILOR'S SEA BOOK, by JAMES GREENWOOD and W. H. ROSSER ; together with Mathematical and Nautical Tables for the Working of the Problems, by HENRY LAW, C.E., and Prof. J. R. YOUNG. Half-bound in leather . . 7/0

"A vast amount of information is contained in this volume, and we fancy in a very short time that it will be seen in the library of almost every ship or yacht afloat."—*Hunt's Yachting Magazine.*

Navigation and Nautical Astronomy,

In Theory and Practice. By Prof. J. R. YOUNG. New Edition. Illustrated . 2/6

"A very complete, thorough, and useful manual for the young navigator."—*Observatory.*

Mathematical Tables,

For Trigonometrical, Astronomical, and Nautical Calculations; to which is prefixed a Treatise on Logarithms, by H. LAW, C.E. Together with a Series of Tables for Navigation and Nautical Astronomy. By Professor J. R. YOUNG. New Edition **4/0**

Masting, Mast-Making, and Rigging of Ships.

Also Tables of Spars, Rigging, Blocks; Chain, Wire, and Hemp Ropes, &c., relative to every class of vessels. By ROBERT KIPPING, N.A. **2/0**

Sails and Sail-Making.

With Draughting, and the Centre of Effort of the Sails. Weights and Sizes of Ropes; Masting, Rigging, and Sails of Steam Vessels, &c. By R. KIPPING, N.A. . **2/6**

Marine Engines and Steam Vessels.

By R. MURRAY, C.E. Eighth Edition, thoroughly Revised, with Additions by the Author and by GEORGE CARLISLE, C.E. **4/6**
"An indispensable manual for the student of marine engineering."—*Liverpool Mercury.*

Iron Ship-Building.

With Practical Examples and Details. By JOHN GRANTHAM. Fifth Edition . **4/0**

Naval Architecture.

An Exposition of the Elementary Principles. By JAMES PEAKE . . . **3/6**

Ships for Ocean and River Service,

Principles of the Construction of. By H. A. SOMMERFELDT **1/6**

An Atlas of Engravings

To Illustrate the above. Twelve large folding Plates. Royal 4to, cloth . . **7/6**

Ships and Boats.

By W. BLAND. Seventh Edition, revised, with numerous Illustrations and Models **1/6**

ARCHITECTURE AND BUILDING.

Constructional Iron and Steel Work,

As applied to Public, Private, and Domestic Buildings. By FRANCIS CAMPIN, C.E. **3/6**
"Anyone who wants a book on ironwork as employed for stanchions, columns, and beams, will find the present volume to be suitable."—*British Architect.*

Building Estates:

A Treatise on the Development, Sale, Purchase, and Management of Building Land. By F. MAITLAND. Second Edition, revised **2/0**
"This book should undoubtedly be added to the library of every professional man dealing with building land."—*Land Agent's Record.*

Science of Building:

An Elementary Treatise on the Principles of Construction. By E. WYNDHAM TARN, M.A. Lond. Third Edition, revised and enlarged **3/6**

Art of Building,

Rudiments of. General Principles of Construction, Strength, and Use of Materials, Working Drawings, Specifications, &c. By EDWARD DOBSON, M.R.I.B.A. &c. **2/0**
"A good book for practical knowledge, and about the best to be obtained."—*Building News.*

Book on Building,

Civil and Ecclesiastical. By Sir EDMUND BECKETT, Bart., LL.D., Q.C., F.R.A.S., Author of "Clocks and Watches and Bells," &c. Second Edition, enlarged. . **4/6**
"A book which is always amusing and nearly always instructive."—*Times.*

Dwelling-Houses,

Erection of, Illustrated by a Perspective View, Plans, Elevations, and Sections of a Pair of Villas, with the Specification, Quantities, and Estimates. By S. H. BROOKS **2/6**

Cottage Building.

By C. BRUCE ALLEN. Eleventh Edition, with Chapter on Economic Cottages for Allotments, by E. E. ALLEN, C.E. **2/0**

Acoustics of Public Buildings:

The Laws of Sound as applied to the Arrangement of Buildings. By Professor T. ROGER SMITH, F.R.I.B.A. New Edition, revised. With numerous Illustrations.
[Just published. **1/6**

Practical Bricklaying.

General Principles of Bricklaying; Arch Drawing, Cutting and Setting; Pointing; Paving, Tiling, &c. By ADAM HAMMOND. With 68 Woodcuts . . . **1/6**
"The young bricklayer will find it infinitely valuable to him."—*Glasgow Herald.*

Art of Practical Brick-Cutting and Setting.

By ADAM HAMMOND. With 90 Engravings **1/6**

Brickwork:

Embodying the General and Higher Principles of Bricklaying, Cutting and Setting; with the Application of Geometry to Roof Tiling, &c. By F. WALKER . . **1/6**
"Contains all that a young tradesman or student needs to learn from books."—*Building News.*

Bricks and Tiles,

Rudimentary Treatise on the Manufacture of. Containing an Outline of the Principles of Brickmaking. By E. DOBSON, M.R.I.B.A. Additions by C. TOMLINSON, F.R.S. Illustrated **3/0**
"The best handbook on the subject. We can safely recommend it as a good investment."—*Builder.*

Practical Brick and Tile Book.

Comprising : BRICK AND TILE MAKING, by E. DOBSON, A.I.C.E. ; Practical BRICK-LAYING, by A. HAMMOND ; BRICK-CUTTING and SETTING, by A. HAMMOND. 550 pp. with 270 Illustrations, strongly half-bound **6/0**

Carpentry and Joinery—

THE ELEMENTARY PRINCIPLES OF CARPENTRY. Chiefly composed from the Standard Work of THOMAS TREDGOLD, C.E. With Additions, and a TREATISE ON JOINERY by E. W. TARN, M.A. Sixth Edition, revised and extended . . **3/6**

Carpentry and Joinery.

Atlas of 35 Plates to accompany and illustrate the foregoing book. With Descriptive Letterpress. 4to. **6/0**
"These two volumes form a complete treasury of carpentry and joinery, and should be in the hands of every carpenter and joiner in the Empire."—*Iron.*

Practical Treatise on Handrailing :

Showing New and Simple Methods. By GEO. COLLINGS. Second Edition, Revised, including a TREATISE ON STAIRBUILDING. With Plates . . **2/6**
"Will be found of practical utility in the execution of this difficult branch of joinery."—*Builder.*

Circular Work in Carpentry and Joinery.

A Practical Treatise on Circular Work of Single and Double Curvature. By GEORGE COLLINGS. Second Edition **2/6**
"Cheap in price, clear in definition, and practical in the examples selected."—*Builder.*

Roof Carpentry :

Practical Lessons in the Framing of Wood Roofs. For the use of Working Carpenters. By GEO. COLLINGS, Author of " Handrailing and Stairbuilding," &c. . **2/0**

Construction of Roofs, of Wood and Iron :

Deduced chiefly from the Works of Robison, Tredgold, and Humber. By E. WYNDHAM TARN, M.A., Architect. Second Edition, revised . . **1/6**
"Mr. Tarn is so thoroughly master of his subject, that although the treatise was founded on the works of others he has given it a distinct value of his own. It will be found valuable by all students."—*Builder.*

The Joints Made and Used by Builders.

By WYVILL J. CHRISTY, Architect. With 160 Woodcuts . . . **3/0**
"The work is deserving of high commendation."—*Builder.*

Shoring,

And its Application : A Handbook for the use of Students. By G. H. BLAGROVE **1/6**
"We recommend this valuable treatise to all students."—*Building News.*

Timber Importer's, Timber Merchant's, and Builder's Standard Guide.

By R. E. GRANDY. **2/0**
"Everything it pretends to be : built up gradually, it leads one from a forest to a treenail, and throws in, as a makeweight, a host of material concerning bricks, columns, cisterns, &c."—*English Mechanic.*

Plumbing :

A Text-Book to the Practice of the Art or Craft of the Plumber. With Chapters upon House Drainage and Ventilation. By WM. PATON BUCHAN, R.P., Sanitary Engineer. Sixth Edition, revised and enlarged, with 380 Illustrations . . . **3/6**
"A text-book which may be safely put into the hands of every young plumber, and which will also be found useful by architects and medical professors."—*Builder.*

Ventilation :

A Text-Book to the Practice of the Art of Ventilating Buildings. By W. P. BUCHAN, R.P., Author of " Plumbing," &c. With 170 Illustrations . . . **3/6**

The Practical Plasterer :

A Compendium of Plain and Ornamental Plaster Work. By WILFRED KEMP **2/0**

House Painting, Graining, Marbling, and Sign Writing :

With a Course of Elementary Drawing, and a Collection of Useful Receipts. By ELLIS A. DAVIDSON. Sixth Edition. Coloured Plates **5/0**
* *The above, in cloth boards, strongly bound, 6/0.*
"A mass of information of use to the amateur and of value to the practical man."—*English Mechanic.*

Grammar of Colouring.

Applied to Decorative Painting and the Arts. By GEORGE FIELD. New Edition, revised and enlarged by ELLIS A. DAVIDSON. With Coloured Plates . . **3/0**

"The book is a most useful *resumé* of the properties of pigments."—*Builder.*

Elementary Decoration:

As Applied to Dwelling-Houses, &c. By JAMES W. FACEY. Illustrated . . **2/0**

"The principles which ought to guide the decoration of dwelling-houses are clearly set forth, and elucidated by examples; while full instructions are given to the learner."—*Scotsman.*

Practical House Decoration.

A Guide to the Art of Ornamental Painting, the Arrangement of Colours in Apartments, and the Principles of Decorative Design. By JAMES W. FACEY . . . **2/6**

. *The last two Works in One handsome Vol., half-bound, entitled* "HOUSE DECORATION, ELEMENTARY AND PRACTICAL," *price* **5/0.**

Warming and Ventilation

Of Domestic and Public Buildings, &c. By C. TOMLINSON, F.R.S. . . **3/0**

Portland Cement for Users.

By HENRY FAIJA, A.M. Inst. C.E. Third Edition, corrected **2/0**

"Supplies in a small compass all that is necessary to be known by users of cement."—*Building News.*

Limes, Cements, Mortars, Concretes, Mastics, Plastering, &c.

By G. R. BURNELL, C.E. Thirteenth Edition **1/6**

Masonry and Stonecutting,

The Principles of Masonic Projection, and their Application to Construction. By E. DOBSON, M.R.I.B.A. **2/6**

Arches, Piers, Buttresses, &c,

Experimental Essays on the Principles of Construction in. By WILLIAM BLAND **1/6**

Quantities and Measurements,

In Bricklayers', Masons', Plasterers', Plumbers', Painters', Paperhangers', Gilders', Smiths', Carpenters' and Joiners' Work. By A. C. BEATON, Surveyor . . **1/6**

"This book is indispensable to builders and their quantity clerks."—*English Mechanic.*

Complete Measurer;

Setting forth the Measurement of Boards, Glass, Timber, and Stone. By R. HORTON. Fifth Edition **4/0**

. *The above, strongly bound in leather, price* **5/0.**

Light :

An Introduction to the Science of Optics. Designed for the Use of Students of Architecture, Engineering, and other Applied Sciences. By E. W. TARN, M.A. . **1/6**

Hints to Young Architects.

By GEORGE WIGHTWICK, Architect, Author of "The Palace of Architecture," &c., &c.. Fifth Edition, revised and enlarged by G. HUSKISSON GUILLAUME, Architect . **3/6**

"A copy ought to be considered as necessary a purchase as a box of instruments."—*Architect.*

Architecture—Orders.

The Orders and their Æsthetic Principles. By W. H. LEEDS. Illustrated . **1/6**

Architecture—Styles.

The History and Description of the Styles of Architecture of Various Countries, from the Earliest to the Present Period. By T. TALBOT BURY, F.R.I.B.A., &c. Illustrated **2/0**

ORDERS AND STYLES OF ARCHITECTURE, *in One Vol.,* **3/6.**

Architecture—Design.

The Principles of Design in Architecture, as deducible from Nature and exemplified in the Works of the Greek and Gothic Architects. By EDW. L. GARBETT, Architect **2/6**

"We know no work that we would sooner recommend to an attentive reader desirous to obtain clear views of the nature of architectural art. The book is a valuable one."—*Builder.*

The three preceding Works in One handsome Vol., half-bound, entitled "MODERN ARCHITECTURE," *price* **6/0.**

Architectural Modelling in Paper,

The Art of. By T. A. RICHARDSON. With Illustrations, engraved by O. JEWITT **1/6**

"A valuable aid to the practice of architectural modelling."—*Builder's Weekly Reporter.*

Perspective for Beginners.

For Students and Amateurs in Architecture, Painting, &c. By G. PYNE . . **2/0**

Glass Staining, and the Art of Painting on Glass.

From the German of Dr. GESSERT and EMANUEL OTTO FROMBERG. With an Appendix on THE ART OF ENAMELLING **2/6**

Vitruvius—The Architecture of Marcus Vitruvius Pollio.

In Ten Books. Translated from the Latin by JOSEPH GWILT, F.S.A., F.R.A.S. With 23 Plates 5/0

N.B.—This is the only Edition of VITRUVIUS *procurable at a moderate price.*

Grecian Architecture,

An Inquiry into the Principles of Beauty in; with an Historical View of the Rise and Progress of the Art in Greece. By the EARL OF ABERDEEN . . . 1/0

** The two preceding Works in One handsome Vol., half-bound, entitled "ANCIENT ARCHITECTURE," price 6/0.

INDUSTRIAL AND USEFUL ARTS.

Cements, Pastes, Glues, and Gums.

A Practical Guide to the Manufacture and Application of the various Agglutinants required for Workshop, Laboratory, or Office Use. With upwards of 900 Recipes and Formulæ. By H. C. STANDAGE 2/0

"As a revelation of what are considered trade secrets, this book will arouse an amount of curiosity among the large number of industries it touches."—*Daily Chronicle.*

Clocks and Watches, and Bells,

A Rudimentary Treatise. By Sir EDMUND BECKETT. Seventh Edition. . 4/6

** The above, handsomely bound, Cloth Boards, 5/6.

"The best work on the subject probably extant. The treatise on bells is undoubtedly the best in the language."—*Engineering.* "The only modern treatise on clock-making."—*Horological Journal.*

Electro-Metallurgy,

Practically Treated. By ALEXANDER WATT. Tenth Edition, enlarged and revised. With Additional Illustrations, and including the most Recent Processes . . 3/6

"From this book both amateur and artisan may learn everything necessary."—*Iron.*

Goldsmith's Handbook,

Containing full Instructions in the Art of Alloying, Melting, Reducing, Colouring, Collecting, and Refining. The processes of Manipulation, Recovery of Waste, Chemical and Physical Properties of Gold ; Solders, Enamels, and other useful Rules and Recipes, &c. By GEORGE E. GEE. Third Edition, considerably enlarged 3/0

"A good, sound, technical educator."—*Horological Journal.*

Silversmith's Handbook,

On the same plan as the above. By GEORGE E. GEE. Second Edition, Revised 3/0

'A valuable sequel to the author's ' Practical Goldworker.'"—*Silversmith's Trade Journal.*

** The two preceding Works, in One handsome Vol., half-bound, entitled "THE GOLDSMITH'S AND SILVERSMITH'S COMPLETE HANDBOOK," 7/0.

Hall-Marking of Jewellery.

Comprising an account of all the different Assay Towns of the United Kingdom ; with the Stamps at present employed ; also the Laws relating to the Standards and Hall-Marks at the various Assay Offices. By GEORGE E. GEE. 3/0

"Deals thoroughly with its subject from a manufacturer's and dealer's point of view."—*Jeweller.*
"A valuable and trustworthy guide."—*English Mechanic.*

Practical Organ Building.

By W. E. DICKSON, M.A., Precentor of Ely Cathedral. Second Edition, Revised 2/6

"The amateur builder will find in this book all that is necessary to enable him personally to construct a perfect organ with his own hands."—*Academy.*

"The best work on the subject that has yet appeared in book form."—*English Mechanic.*

Coach-Building :

A Practical Treatise, Historical and Descriptive. By JAMES W. BURGESS . 2/6

"This handbook will supply a long-felt want, not only to manufacturers themselves, but more particularly apprentices, and others whose occupations may be in any way connected with the trade of coach-building."—*European Mail.*

Brass Founder's Manual :

Instructions for Modelling, Pattern Making, Moulding, Turning, &c. By W. GRAHAM 2/0

Sheet Metal-Worker's Guide.

A Practical Handbook for Tinsmiths, Coppersmiths, Zincworkers, &c., with 46 Diagrams and Working Patterns. By W. J. E. CRANE. Second Edition, revised 1/6

"The author has acquitted himself with considerable tact in choosing his examples, and with no less ability in treating them."—*Plumber.*

Sewing Machinery.

Its Construction, History, &c. With full Technical Directions for Adjusting, &c. By J. W. URQUHART, C.E. 2/0

"A full description of the principles and construction of the leading machines, and minute instructions as to their management."—*Scotsman.*

Gas Fitting:
A Practical Handbook. By JOHN BLACK. Revised Edition. With 130 Illustrations **2/6**
"Contains all the requisite information for the successful fitting of houses with a gas service, &c. It is written in a simple practical style, and we heartily recommend it."—*Plumber and Decorator.*

Construction of Door Locks.
From the Papers of A. C. HOBBS. Edited by CHARLES TOMLINSON, F.R.S. With a Note upon IRON SAFES by ROBERT MALLET. Illustrated. **2/6**

The Model Locomotive Engineer, Fireman, and Engine-Boy.
Comprising an Historical Notice of the Pioneer Locomotive Engines and their Inventors. By MICHAEL REYNOLDS **3/6**

Art of Letter Painting made Easy.
By JAMES GREIG BADENOCH. With 12 full-page Engravings of Examples. . **1/6**
"Any intelligent lad who fails to turn out decent work after studying this system, has mistaken his vocation."—*English Mechanic.*

Art of Boot and Shoemaking,
Including Measurement, Last-fitting, Cutting-out, Closing and Making; with a Description of the most Approved Machinery employed. By J. B. LENO. . . **2/0**
"By far the best work ever written on the subject. The chapter on clicking, which shows how waste may be prevented, will save fifty times the price of the book."—*Scottish Leather Trader.*

Mechanical Dentistry:
A Practical Treatise on the Construction of the Various Kinds of Artificial Dentures, comprising also Useful Formulæ, Tables and Receipts. By C. HUNTER . . **3/0**

Wood Engraving:
A Practical and Easy Introduction to the Study of the Art. By W. N. BROWN **1/6**

Laundry Management.
A Handbook for Use in Private and Public Laundries. Including Accounts of Modern Machinery and Appliances. By the EDITOR of *The Laundry Journal* . . **2/0**
"This book should certainly occupy an honoured place on the shelves of all housekeepers who wish to keep themselves *au courant* of the newest appliances and methods."—*The Queen.*

AGRICULTURE, GARDENING, ETC.

Draining and Embanking.
A Practical Treatise. By JOHN SCOTT, late Professor of Agriculture and Rural Economy at the Royal Agricultural College, Cirencester. With 68 Illustrations. . . . **1/6**
"A valuable handbook to the engineer as well as to the surveyor."—*Land.*

Irrigation and Water Supply:
A Practical Treatise on Water Meadows, Sewage Irrigation, Warping, &c.; on the Construction of Wells, Ponds, and Reservoirs, &c. By Professor J. SCOTT . **1/6**
"A valuable and indispensable book for the estate manager and owner."—*Forestry.*

Farm Roads, Fences, and Gates:
A Practical Treatise on the Roads, Tramways, and Waterways of the Farm; the Principles of Enclosures; and on Fences, Gates, and Stiles. By Prof. JOHN SCOTT . **1/6**
"A useful practical work, which should be in the hands of every farmer."—*Farmer.*

Farm Buildings:
A Practical Treatise on the Buildings necessary for various kinds of Farms, their Arrangement and Construction, with Plans and Estimates. By Professor J. SCOTT **2/0**
"No one who is called upon to design farm buildings can afford to be without this work."—*Builder.*

Barn Implements and Machines:
Treating of the Application of Power to the Operations of Agriculture; and of the various Machines used in the Threshing-barn, in the Stockyard, Dairy, &c. By Professor JOHN SCOTT. With 123 Illustrations **2/0**

Field Implements and Machines:
With Principles and Details of Construction and Points of Excellence, their Management, &c. By Professor JOHN SCOTT. With 138 Illustrations **2/0**

Agricultural Surveying:
A Treatise on Land Surveying, Levelling, and Setting-out; with Directions for Valuing and Reporting on Farms and Estates. By Professor J. SCOTT . . **1/6**

Farm Engineering.
By Professor JOHN SCOTT, Comprising the above Seven Volumes in One, 1,150 pages, and over 600 Illustrations. Half-bound **12/0**
"A copy of this work should be treasured up in every library where the owner thereof is in any way connected with land."—*Farm and Home.*

Outlines of Farm Management.

Treating of the General Work of the Farm; Stock; Contract Work; Labour, &c. By R. Scott Burn, Author of "Outlines of Modern Farming," &c. **2/6**

"The book is eminently practical, and may be studied with advantage by beginners in agriculture, while it contains hints which will be useful to old and successful farmers."—*Scotsman.*

Outlines of Landed Estates Management.

Treating of the Varieties of Lands, Methods of Farming, the Setting-out of Farms, &c.; Roads, Fences, Gates, Irrigation, Drainage, &c. By R. S. Burn **2/6**

"A complete and comprehensive outline of the duties appertaining to the management of landed estates."—*Journal of Forestry.*

The above Two Vols. in One, handsomely half-bound, entitled "Cutlines of Landed Estates and Farm Management." By Robert Scott Burn. *Price* **6/0.**

Soils, Manures, and Crops.

(Vol. I. Outlines of Modern Farming.) By R. Scott Burn. Woodcuts . **2/0**

Farming and Farming Economy,

Historical and Practical. (Vol. II. Outlines of Modern Farming.) By R. Scott Burn **3/0**

"Eminently calculated to enlighten the agricultural community on the varied subjects of which it treats; hence it should find a place in every farmer's library."—*City Press.*

Stock: Cattle, Sheep, and Horses.

(Vol. III. Outlines of Modern Farming.) By R. Scott Burn. Woodcuts **2/6**

"The author's grasp of his subject is thorough, and his grouping of facts effective. . . . We commend this excellent treatise."—*Weekly Dispatch.*

Dairy, Pigs, and Poultry.

(Vol. IV. Outlines of Modern Farming.) By R. Scott Burn. Woodcuts **2/0**

"We can testify to the clearness and intelligibility of the matter, which has been compiled from the best authorities."—*London Review.*

Utilization of Sewage, Irrigation, &c.

(Vol. V. Outlines of Modern Farming.) By R. Scott Burn. Woodcuts **2/6**

"A work containing valuable information, which will recommend itself to all interested in modern farming."—*Field.*

Outlines of Modern Farming.

By R. Scott Burn, Author of "Landed Estates Management," &c. Consisting of the above Five Volumes in One, 1,250 pp., profusely Illustrated, half-bound . **12/0**

"The aim of the author has been to make his work at once comprehensive and trustworthy, and in this aim he has succeeded to a degree which entitles him to much credit."—*Morning Advertiser.*

Book-keeping for Farmers and Estate Owners.

A Practical Treatise, presenting, in Three Plans, a System adapted for all classes of Farms. By J. M. Woodman, Chartered Accountant. Third Edition, revised . **2/6**

*** *The above, in cloth boards, strongly bound,* **3/6.**

"Will be found of great assistance by those who intend to commence a system of book-keeping, the author's examples being clear and explicit, and his explanations full and accurate."—*Live Stock Journal.*

Ready Reckoner for Admeasurement of Land,

By A. Arman. Third Edition, revised and extended by C. Norris, Surveyor . **2/0**

"A very useful book to all who have land to measure."—*Mark Lane Express.*
"Should be in the hands of all persons having any connection with land."—*Irish Farm.*

Ready Reckoner for Millers, Corn Merchants,

And Farmers. Second Edition, revised, with a Price List of Modern Flour Mill Machinery. By W. S. Hutton, C.E. **2/0**

"Will prove an indispensable *vade mecum.* Nothing has been spared to make the book complete and perfectly adapted to its special purpose."—*Miller.*

The Hay and Straw Measurer:

New Tables for the use of Auctioneers, Valuers, Farmers, Hay and Straw Dealers, &c., forming a complete Calculator and Ready Reckoner. By John Steele . **2/0**

"A most useful handbook. It should be in every professional office where agricultural valuations are conducted."—*Land Agent's Record.*

Meat Production:

A Manual for Producers, Distributors, and Consumers of Butchers' Meat. By John Ewart **2/6**

"A compact and handy volume on the meat question."—*Meat and Provision Trades Review.*

Sheep:

The History, Structure, Economy, and Diseases of. By W. C. Spooner. Fifth Edition, with Engravings, including Specimens of New and Improved Breeds . . . 3/6
"The book is decidedly the best of the kind in our language."—*Scotsman.*

Market and Kitchen Gardening.

By C. W. Shaw, late Editor of "Gardening Illustrated" 3/0
"The most valuable compendium of kitchen and market-garden work published."—*Farmer.*

Kitchen Gardening made Easy.

Showing the best means of Cultivating every known Vegetable and Herb, &c., with directions for management all the year round. By Geo. M. F. Glenny. Illustrated 1/6
"This book will be found trustworthy and useful."—*North British Agriculturist.*

Cottage Gardening;

Or, Flowers, Fruits, and Vegetables for Small Gardens. By E. Hobday . . 1/6
"Definite instructions as to the cultivation of small gardens."—*Scotsman.*

Garden Receipts.

Edited by Charles W. Quin. Third Edition 1/6
"A singularly complete collection of the principal receipts needed by gardeners."—*Farmer.*

Fruit Trees,

The Scientific and Profitable Culture of. From the French of M. Du Breuil. Fourth Edition, carefully Revised by George Glenny. With 187 Woodcuts . . 3/6
"The book teaches how to prune and train fruit trees to perfection."—*Field.*

Tree Planter and Plant Propagator:

With numerous Illustrations of Grafting, Layering, Budding, Implements, Houses, Pits, &c. By S. Wood, Author of "Good Gardening," &c. 2/0
"Sound in its teaching and very comprehensive in its aim. It is a good book."—*Gardeners' Magazine.*

Tree Pruner:

Being a Practical Manual on the Pruning of Fruit Trees, including also their Training and Renovation, also treating of the Pruning of Shrubs, Climbers, and Flowering Plants. With numerous Illustrations. By Samuel Wood, Author of "Good Gardening," &c. 1/6
"A useful book, written by one who has had great experience."—*Mark Lane Express.*

*** *The above Two Vols. in One, handsomely half-bound, entitled* "The Tree Planter, Propagator and Pruner." *By* Samuel Wood. *Price* 3/6.

Art of Grafting and Budding.

By Charles Baltet. With Illustrations 2/6
"The one standard work on this subject."—*Scotsman.*

MATHEMATICS, ARITHMETIC, ETC.

Descriptive Geometry,

An Elementary Treatise on; with a Theory of Shadows and of Perspective, extracted from the French of G. Monge. To which is added a Description of the Principles and Practice of Isometrical Projection. By J. F. Heather, M.A. With 14 Plates. 2/0

Practical Plane Geometry:

Giving the Simplest Modes of Constructing Figures contained in one Plane and Geometrical Construction of the Ground. By J. F. Heather, M.A. . . 2/0
"The author is well-known as an experienced professor, and the volume contains as complete a collection of problems as is likely to be required in ordinary practice."—*Architect.*

Analytical Geometry and Conic Sections.

By James Hann. New Edition, Enlarged by Professor J. R. Young . . 2/0
"The author's style is exceedingly clear and simple, and the book is well adapted for the beginner and those who may be obliged to have recourse to self-tuition."—*Engineer.*

Euclid,

The Elements of; with many Additional Propositions and Explanatory Notes; to which is prefixed an Introductory Essay on Logic. By Henry Law, C.E. . 2/6
*** *Sold also separately, viz.* :—
Euclid. The First Three Books. By Henry Law, C.E. 1/6
Euclid. Books 4, 5, 6, 11, 12. By Henry Law, C.E. 1/6

Plane Trigonometry,

The Elements of. By James Hann, M.A. Sixth Edition 1/6

Spherical Trigonometry,
The Elements of. By JAMES HANN. Revised by CHARLES H. DOWLING, C.E. **1/0**
  ₊ *Or with " The Elements of Plane Trigonometry," in One Vol.,* **2/6**.

Differential Calculus,
Elements of the. By W. S. B. WOOLHOUSE, F.R.A.S., &c. **1/6**

Integral Calculus.
By HOMERSHAM COX, B.A. . : . . : . : **1/0**

Algebra,
The Elements of. By JAMES HADDON, M.A., formerly Mathematical Master of King's College School. With Appendix, containing Miscellaneous Investigations, and a collection of Problems **2/0**

Key and Companion to the Above.
An extensive repository of Solved Examples and Problems in Illustration of the various Expedients necessary in Algebraical Operations. By J. R. YOUNG. . **1/6**

Commercial Book-keeping.
With Commercial Phrases and Forms in English, French, Italian, and German. By JAMES HADDON, M.A., formerly Mathematical Master, King's College School . **1/6**

Arithmetic,
A Rudimentary Treatise on : with full Explanations of its Theoretical Principles, and numerous Examples for Practice. For the use of Schools and for Self-Instruction. By J. R. YOUNG, late Professor of Mathematics in Belfast College. Eleventh Edition **1/6**

Key to the Above.
By J. R. YOUNG . . . : . ; ; ; ; **1/6**

Equational Arithmetic,
Applied to Questions of Interest, Annuities, Life Assurance, and General Commerce ; with various Tables by which all calculations may be greatly facilitated. By W. HIPSLEY **2/0**

Arithmetic,
Rudimentary, for the Use of Schools and Self-Instruction. By JAMES HADDON, M.A. Revised by ABRAHAM ARMAN **1/6**

Key to the Above.
By A. ARMAN . . . . : . : : . **1/6**

Mathematical Instruments,
A Treatise on ; Their Construction, Adjustment, Testing, and Use concisely explained. By J. F. HEATHER, M.A., of the Royal Military Academy, Woolwich. Fourteenth Edition, Revised with Additions, by A. T. WALMISLEY, M.I.C.E., Fellow of the Surveyors' Institution. Original Edition in One Vol., Illustrated . . . **2/0**
₊ *In ordering be careful to say " Original Edition," to distinguish it from the Enlarged Edition in Three Vols. (see below).*

Drawing and Measuring Instruments.
Including—I. Instruments employed in Geometrical and Mechanical Drawing, and in the Construction, Copying, and Measurement of Maps and Plans. II. Instruments used for the purposes of Accurate Measurement, and for Arithmetical Computations. By J. F. HEATHER, M.A. **1/6**

Optical Instruments.
Including (more especially) Telescopes, Microscopes, and Apparatus for producing copies of Maps and Plans by Photography. By J. F. HEATHER, M.A. Illustrated **1/6**

Surveying and Astronomical Instruments.
Including—I. Instruments used for Determining the Geometrical Features of a portion of Ground. II. Instruments employed in Astronomical Observations. By J. F. HEATHER, M.A. Illustrated **1/6**
₊ *The above Three Volumes form an enlargement of the Author's original work, " Mathematical Instruments," price* **2/0**.

Mathematical Instruments:
Their Construction, Adjustment, Testing and Use. Comprising Drawing, Measuring, Optical, Surveying, and Astronomical Instruments. By J. F. HEATHER, M.A. Enlarged Edition, for the most part re-written. Three Parts as above . . **4/6**
"An exhaustive treatise, belonging to the well-known Weale's Series. Mr. Heather's experience well qualifies him for the task he has so ably fulfilled."—*Engineering and Building Times.*

Slide Rule, and How to Use It.
Containing full, easy, and simple Instructions to perform all Business Calculations with unexampled rapidity and accuracy. By CHARLES HOARE, C.E. With a Slide Rule, in tuck of cover. Fifth Edition **2/6**

Mathematical Tables,

For Trigonometrical, Astronomical, and Nautical Calculations ; to which is prefixed a Treatise on Logarithms. by H. LAW, C.E. Together with a Series of Tables for Navigation and Nautical Astronomy. By Professor J. R. YOUNG. New Edition **4/0**

Logarithms.

With Mathematical Tables for Trigonometrical, Astronomical, and Nautical Calculations. By HENRY LAW, C.E. Revised Edition. (Forming part of the above work) **3/0**

Theory of Compound Interest and Annuities:

With Tables of Logarithms for the more Difficult Computations of Interest, Discount, Annuities, &c., in all their Applications and Uses for Mercantile and State Purposes. By FEDOR THOMAN, of the Société Crédit, Mobilier, Paris. Fourth Edition . **4/0**

"A very powerful work, and the author has a very remarkable command of his subject."--Professor A. DE MORGAN. "We recommend it to the notice of actuaries and accountants."--*Athenæum.*

Treatise on Mathematics,

As applied to the Constructive Arts. By FRANCIS CAMPIN, C.E., &c. 2nd Edn. **3/0**

"Should be in the hands of everyone connected with building construction."--*Builder's Reporter.*

Astronomy.

By the late Rev. ROBERT MAIN, M.A., F.R.S. Third Edition revised and corrected. By WILLIAM THYNNE LYNN, B.A., F.R.A.S. **2/0**

"A sound and simple treatise, very carefully edited and a capital book for beginners."--*Knowledge*

Statics and Dynamics,

The Principles and Practice of ; embracing also a clear development of Hydrostatics, Hydrodynamics and Central Forces. By T. BAKER, C.E. Fourth Edition . **1/6**

BOOKS OF REFERENCE AND
MISCELLANEOUS VOLUMES.

Manual of the Mollusca :

A Treatise on Recent and Fossil Shells. By Dr. S. P. WOODWARD, A.L.S. With Appendix by RALPH TATE, A.L.S., F.G.S. With numerous Plates and 300 Woodcuts **7/6**

"A storehouse of conchological and geological information."--*Hardwicke's Science Gossip.*

Dictionary of Painters,

And Handbook for Picture Amateurs ; being a Guide for Visitors to Public and Private Picture Galleries, and for Art Students, including Glossary of Terms, &c. By PHILIPPE DARYL, B.A. **2/6**

"Considering its small compass, really admirable. We cordially recommend the book."--*Builder.*

Painting Popularly Explained.

By THOMAS JOHN GULLICK, Painter, and JOHN TIMBS, F.S.A. Including Fresco, Oil, Mosaic, Water Colour, Water-Glass, Tempera, Encaustic, Miniature, Painting on Ivory, Vellum, Pottery, Enamel, Glass, &c. Fifth Edition **5/0**

. *Adopted as a Prize Book at South Kensington.*

"Much may be learned, even by those who fancy they do not require to be taught, from the careful perusal of this unpretending but comprehensive treatise."--*Art Journal.*

Dictionary of Terms used in Architecture,

Building, Engineering, Mining, Metallurgy, Archæology, the Fine Arts, &c. By JOHN WEALE. Sixth Edition. Edited by ROBT. HUNT, F.R.S., Keeper of Mining Records, Editor of "Ure's Dictionary." Numerous Illustrations . **5/0**

. *The above, strongly bound in cloth boards, price 6/0.*

"The best small technological dictionary in the language."--*Architect.*

Music,

A Rudimentary and Practical Treatise on. By CHARLES CHILD SPENCER . **2/6**

"Mr. Spencer has marshalled his information with much skill, and yet with a simplicity that must recommend his works to all who wish to thoroughly understand music."--*Weekly Times.*

Pianoforte,

The Art of Playing the. With Exercises and Lessons. By C. C. SPENCER . **1/6**

"A sound and excellent work, written with spirit, and calculated to inspire the pupil with a desire to aim at high accomplishment in the art."--*School Board Chronicle*

House Manager:

Being a Guide to Housekeeping, Practical Cookery, Pickling and Preserving, Household Work, Dairy Management, the Table and Dessert, Cellarage of Wines, Homebrewing and Wine-making, the Boudoir and Dressing-room, Travelling, Stable Economy, Gardening Operations, &c. By AN OLD HOUSEKEEPER . . **3/6**

" We find here directions to be discovered in no other book, tending to save expense to the pocket, as well as labour to the head."—*John Bull.*

Manual of Domestic Medicine.

By R. GOODING, B.A., M.D. Intended as a Family Guide in all Cases of Accident and Emergency. Third Edition, carefully revised **2/0**

"The author has, we think, performed a useful service by placing at the disposal of those situated, by unavoidable circumstances, at a distance from medical aid, a reliable and sensible work in which professional knowledge and accuracy have been well seconded by the ability to express himself in ordinary untechnical language."—*Public Health.*

Management of Health.

A Manual of Home and Personal Hygiene. By the Rev. JAMES BAIRD, B.A. **1/0**

"The author gives sound instruction for the preservation of health."—*Athenæum.*

"It is wonderfully reliable, it is written with excellent taste, and there is instruction crowded into every page."—*English Mechanic.*

House Book,

Comprising : I. THE HOUSE MANAGER. By AN OLD HOUSEKEEPER. II. DOMESTIC MEDICINE. By RALPH GOODING, M.D. III. MANAGEMENT OF HEALTH. By JAMES BAIRD. In One Vol., strongly half-bound **6/0**

Natural Philosophy,

For the Use of Beginners. By CHARLES TOMLINSON, F.R.S. **1/6**

Electric Telegraph :

Its History and Progress ; with Descriptions of some of the Apparatus. By R. SABINE, C.E., F.S.A., &c. **3/0**

" Essentially a practical and instructive work."—*Daily Telegraph.*

Handbook of Field Fortification.

By Major W. W. KNOLLYS, F.R.G.S. With 163 Woodcuts. **3/0**

"A well-timed and able contribution to our military literature. . . . The author supplies, in clear business style, all the information likely to be practically useful."—*Chambers of Commerce Chronicle.*

Logic,

Pure and Applied. By S. H. EMMENS. Third Edition. **1/6**

" This admirable work should be a text-book not only for schools, students, and philosophers, for all *litterateurs* and men of science, but for those concerned in the practical affairs of life, &c."—*The News.*

Locke's Essays on the Human Understanding.

Selections, with Notes by S. H. EMMENS **1/6**

Compendious Calculator

(*Intuitive Calculations*) ; or Easy and Concise Methods of performing the various Arithmetical Operations required in Commercial and Business Transactions ; together with Useful Tables, &c. By DANIEL O'GORMAN. Twenty-seventh Edition, carefully revised by C. NORRIS **2/6**

** *The above, strongly half-bound, price 3/6.*

" It would be difficult to exaggerate the usefulness of this book to everyone engaged in commerce or manufacturing industry. It is crammed full with rules and formulæ for shortening and employing calculations in money, weights and measures, &c., of every sort and description."—*Knowledge.*

Measures, Weights, and Moneys of all Nations,

And an Analysis of the Christian, Hebrew, and Mahometan Calendars. By W. S. B. WOOLHOUSE, F.R.A.S., F.S.S. Seventh Edition **2/6**

"A work necessary for every mercantile office."—*Building Trades Journal.*

Grammar of the English Tongue,

Spoken and Written. With an Introduction to the Study of Comparative Philology. By HYDE CLARKE, D.C.L. Fifth Edition **1/6**

Dictionary of the English Language,

As Spoken and Written. Containing about 100,000 Words. By HYDE CLARKE, D.C.L. **3/6**

Complete with the GRAMMAR, *5/6.*

Composition and Punctuation,

Familiarly Explained for those who have neglected the Study of Grammar. By JUSTIN BRENAN. Eighteenth Edition **1/6**

French Grammar.

With Complete and Concise Rules on the Genders of French Nouns. By
G. L. STRAUSS, Ph.D. **1/6**

English-French Dictionary.

By ALFRED ELWES **2/0**

French Dictionary.

In Two Parts: I. French-English. II. English-French. Complete in One Vol. **3/0**
.·. *Or with the* GRAMMAR, **4/6.**

French and English Phrase Book.

Containing Introductory Lessons, with Translations, Vocabularies of Words, Collection
of Phrases, and Easy Familiar Dialogues **1/6**

German Grammar.

Adapted for English Students, from Heyse's Theoretical and Practical Grammar, by
Dr. G. L. STRAUSS **1/6**

German Triglot Dictionary.

By N. E. S. A. HAMILTON. Part I. German-French-English. Part II. English-
German-French. Part III. French-German-English **3/0**

German Triglot Dictionary

(As above). Together with German Grammar in One Vol. **5/0**

Italian Grammar

Arranged in Twenty Lessons, with Exercises. By ALFRED ELWES . . . **1/6**

Italian Triglot Dictionary,

Wherein the Genders of all the Italian and French Nouns are carefully noted down.
By ALFRED ELWES. Vol I. Italian-English-French **2/6**

Italian Triglot Dictionary.

By ALFRED ELWES. Vol. II. English-French-Italian **2/6**

Italian Triglot Dictionary.

By ALFRED ELWES. Vol. III. French-Italian-English **2/6**

Italian Triglot Dictionary

(As above). In One Vol. **7/6**

Spanish Grammar.

In a Simple and Practical Form. With Exercises. By ALFRED ELWES . . **1/6**

Spanish-English and English-Spanish Dictionary.

Including a large number of Technical Terms used in Mining, Engineering, &c., with
the proper Accents and the Gender of every Noun. By ALFRED ELWES . . **4/0**
.·. *Or with the* GRAMMAR, **6/0.**

Portuguese Grammar,

In a Simple and Practical Form. With Exercises. By ALFRED ELWES. . **1/6**

Portuguese English and English-Portuguese Dictionary.

Including a large number of Technical Terms used in Mining, Engineering, &c., with
the proper Accents and the Gender of every Noun. By ALFRED ELWES. Third
Edition, revised· **5/0**
.·. *Or with the* GRAMMAR, **7/0.**

Animal Physics,

Handbook of. By DIONYSIUS LARDNER, D.C.L. With 520 Illustrations. In One Vol.
(732 pages), cloth boards **7/6**
.·. *Sold also in Two Parts, as follows:—*

ANIMAL PHYSICS. By Dr. LARDNER. Part I., Chapters I.—VII. . . **4/0**
ANIMAL PHYSICS. By Dr. LARDNER. Part II., Chapters VIII.—XVIII. . **3/0**

INDEX TO CATALOGUE.